CW01431093

Quarterly Essay

CONTENTS

Quarterly Essay is published four times a year by Black Inc., an imprint of Schwartz Publishing Pty Ltd
Publisher: Morry Schwartz

ISBN 1 86395 339 6

Subscriptions (4 issues): $46.95 a year within Australia incl. GST (Institutional subs. $52.95). Outside Australia $74.95. Payment may be made by Mastercard, Visa or Bankcard, or by cheque made out to Schwartz Publishing. Payment includes postage and handling.

To subscribe, fill out and post the subscription form on the last page of this essay, or subscribe online at:

www.quarterlyessay.com

Correspondence and subscriptions should be addressed to the Editor at:
Black Inc.
Level 5, 289 Flinders Lane
Melbourne VIC 3000 Australia
Phone: 61 3 9654 2000
Fax: 61 3 9654 2290
Email: quarterlyessay@blackincbooks.com
http://www.quarterlyessay.com

Editor: Peter Craven
Management: Silvia Kwon
Managing Editor: Chris Feik
Production Co-ordinator: Sophy Williams
Publicity: Meredith Kelly
Design: Guy Mirabella
Printer: Griffin Press

Quarterly Essay aims to present significant contributions to political, intellectual and cultural debate. It is a magazine in extended pamphlet form and by publishing in each issue a single writer at a length of at least 20,000 words we hope to mediate between the limitations of the newspaper column, where there is the danger that evidence and argument can be swallowed up by the form, and the kind of full-length study of a subject where the only readership is a necessarily specialised one. *Quarterly Essay* aims for the attention of the committed general reader. Although it is a periodical which wants subscribers, each number of the journal is the length of a short book because we want our writers to have the opportunity to speak to the broadest possible audience without condescension or populist short-cuts. *Quarterly Essay* wants to get away from the tyranny that space limits impose in contemporary journalism and we give our essayists the space to express the evidence for their views and those who disagree with them the chance to reply at whatever length is necessary. *Quarterly Essay* will not be confined to politics but is centrally concerned with it. We are not interested in occupying any particular point on the political map and we hope to bring our readership the widest range of political and cultural opinion which is compatible with truth-telling, style and command of the essay form.

INTRODUCTION

Beautiful lies: the famous Mark Twain remark runs deep in the Australian consciousness and it's not hard to see why this nation would have been flattered to have the author of *Huckleberry Finn* and the chronicler of life on the Mississippi taking this view of its convicts and its goldfields and its explorers. After all, Mark Twain was the bard of a new world America that turned its back on all that New Englandism and he bequeathed to it the vigour of a new vernacular, the thing, as Hemingway said, out of which all subsequent American literature came.

Twain was a pioneering writer and an anti-traditionalist. For Tim Flannery the tradition in the Australian handling of the environment is one of mendacity and self-delusion, a betrayal of the truth that has as its logical correlative a form of colonial subservience. The Snowy Mountains Scheme, that flagship of post-war immigration, was just another piece of Australian self-delusion that had a ruinous effect on the river system and filled a naive country with self-congratulation about its post-war openness to non-British immigrants even though the real fear was of the Asian invader.

For Tim Flannery we are always kidding ourselves and we have been doing so ever since the humanistic and Enlightenment openness of the First Fleeters like Watkin Tench gave way to the narrow Social Darwinism of the Victorians, and Australia turned its back on anything it might have learnt from the Aborigines. Far from being nobody's country, the Australia the Aborigines had constructed was a subtle pattern of sustainability in which the hunters and the food gatherers had an intimate knowledge of the land and the way its bush needed to be burnt off if fire was not to ravage all before it.

It is with the mass of settlers (and their derived hierarchies) that the environment becomes imperilled: the fires rage, the rivers turn salty, the richness of plant and bird and animal life, all that diversity, starts to fade like a flower that has been plucked to no purpose. Tim Flannery asks us to look at what nineteenth-century Melbourne did to the Yarra River, which turned into an open drain, its waterfowl so many birds of yester-year, for the next century and more.

Tim Flannery is an unusual Australian zoologist because he is a poly-math who is as steeped in history and literature as he is in science. That's one reason why Watkin Tench is one of his heroes: Tench is the person who articulates with grace and humour the very moving moment when Aborigine and British Australian meet over the cultural gap of thousands of years and know what's what: a worried look is still a worried look, a sexual come on is still a sexual come on.

Flannery is at pains to emphasise that early "convict" Australia was a richer thing than its colonial successor: it could embrace all manner of difference including, at one point, a Muslim festival.

But the subsequent society (colonial in Flannery's judgement until yes-terday at best) became timid and mendacious and it is difficult to know how to kick against the lies in a context of distorted truths. What are we to make of the historians and anthropologists who disingenuously denied the existence of Aboriginal cannibalism, even if that cannibalism was sacramental in nature?

What, for that matter, Flannery asks, are we to make of the myth that multiculturalism and immigration brought us richer and more variegated food?

It is essential to Tim Flannery's stance that he wants to ask questions like this – so much so that Lionel Trilling's phrase for E.M. Forster, "a lib-eral at war with the liberal imagination", can seem to fit him – more particularly if we interpret imagination negatively to include a distorted vision of the truth.

Tim Flannery is the kind of environmentalist who wants to point out that grand river schemes can drown rivers as well as create parasitic levels of agricultural dependence and hideous problems of salinity. He is also a zoologist and a man who loves animals who is bemused by the fact that people who want to look after koalas and save kangaroos – despite the importance of the kangaroo industry – are likely to attract more support, both material and emotive, than those who have a coherent policy for preserving the biodiversity of the country.

And wilderness campaigns (though there was every reason for cheering them on) have been no salvation. The species die off in the national parks as well.

Tim Flannery is a man who wants to cleanse the map of the wrong kind of myth. It irritates him that there are campaigns to save the wrong – dumb – variety of whale just as it irritates him that so much do-goodism and romantic sentimentality has been invested in notions of wilderness that present green consciousness as a kind of end in itself.

He is terrified by the fact of global warming and fears that our grandchildren will endure the kinds of shifts in temperature previous generations only saw over 5,000 years.

He is intensely political about these matters even though his temperament is the opposite of doctrinaire. He says, in a characteristically trenchant formulation, that it is its failure to sign the Kyoto Protocol that history will judge the Howard government most harshly for, not the *Tampa* or the refugee crisis or the war in Iraq.

And the question of population, in both its long term and its political aspects, is part of the subject of Tim Flannery's story.

If the CSIRO's calculations are right then we are living in a fools' paradise that cannot admit of long continuance without a radical drop in quality of life. The promises of prime ministers about a massively increased population represent a rhetoric of folly which is in no way helped by the mindless hope that the future and its technology will find a way to cope with every exigency; indeed with everything that gives cogency to today's dire prophecies.

To Tim Flannery this is the hopefulness of a nation of liars. It also has pertinence to his scepticism about both Australian tolerance and Australia's ability, in terms of resources, to act on its own would-be generosity. Every period of Australian history after the first one has been characterised by a xenophobia that has proved irrational, and yet the xenophobia nonetheless persists in mutated form.

In the latter part of this *Quarterly Essay* Tim Flannery addresses the question of what Australia can do for the refugees who have been displaced partly as a result of this country's collusion with American foreign policy. He does not shy away from the moral obligations that face us, nor – as a man intent on the future of our environment – does he pretend that these obligations can be met easily in a world of dwindling resources.

Beautiful Lies is a challenging *Quarterly Essay* because Tim Flannery has the kind of mind that cannot abide the facile prospect of a hypothetical solution some time in the future. This essay is written as a thundering no to the characteristic Australian assumption that "She'll be right."

For Tim Flannery there is no rational basis for believing that she or it will be. This is a *Quarterly Essay* written in the passionate belief that we need a coherent policy on population and the period of self-examination such a policy would pre-suppose. If we do not have one we will never be in a position to do justice to our pieties about the dispossessed people of the earth; indeed our children's children will have every reason to think that we have dishonoured their birthright by living a lie for so long that the riches of the Australian earth have slipped from our hands.

Although *Beautiful Lies* is in its implications a deeply pessimistic essay it is also a political intervention born of a fervent sense of what has to be done. It is an historical essay which advocates that we should fall back on our best traditions, in particular the humanistic one that allowed Tench and his comrades to see in the Aborigines, in Bennelong and the rest of them, fellow human beings with the same desires and expressiveness as ourselves, who had learnt a good deal about the land they lived in.

For Tim Flannery Australia can never be nobody's country; on the contrary, it is a country that everybody from the Aborigines to the refugees has a claim on. What we must do is make the greatest effort we can to sustain what we have and share it justly.

Peter Craven

| BEAUTIFUL LIES | Population and Environment in Australia |

Tim Flannery

On 17 October 1999, thousands of ageing workers gathered together to celebrate the fiftieth anniversary of the inauguration of the Snowy Mountains Scheme. Television documentaries showed elderly men who had come from eastern Europe reminiscing about arriving in Australia to work in the Snowy Mountains, how audacious the scheme's engineering was, and how they coped with life in the insular Australia of the long-ago 1950s. Their reminiscences struck a chord with the many Australians who owed their citizenship and a new start in life to the scheme, or who had come to the country at around the same time. They had left a Europe that had been ravaged by war to become part of what they were told was a noble enterprise.

But at the very time those documentaries were going to air, the Snowy scheme was starting to be seen in a quite different light. It was, many now agree, an environmental catastrophe, and the workers who built it were the innocent perpetrators of a national tragedy. While Victorian fisheries once fed by the Snowy River languished, the water that might have saved them was at work leaching salt from the rock and depositing it into the floodplain and the waters of the Murray, contributing greatly to one of the most serious environmental crises ever to hit Australia – salination.

The Snowy project was in large part created precisely to attract and give meaningful work to migrants. After the war with Japan, Australia had such a fear of Asia that it sought to boost its population with immigration from Europe at almost any cost. Had post-war immigration not occurred in Australia, and had fertility stayed at 1930s–early 1940s levels, Australia's population today would stand at 7.6 million, after reaching a peak of 7.8 million in 1968. The Snowy scheme was part of the cost of population increase, for it helped prevent social upheaval and resentment by providing work for the new migrants and a noble *raison d'être* for their influx. The scheme was built on a distinctive xenophobia – fear of the yellow peril – and it appealed to one of the most cherished myths of the Anglo-Celtic Australians, the dream of turning the rivers inland. It was a sure-fire winner for the governments, state and federal, of the day. Yet the unintended consequences of this grand and expedient enterprise have come back so relentlessly that the hubristic nature of the project could not be more evident.

You would have hoped that the old nexus between unsustainable dreams, environmental damage and population growth had been left behind in the 1950s. Yet today we still hear of schemes for turning the rivers inland; and Australia still lacks a population policy. It seems that even now, when it is so clear that our environment is in peril, we remain incapable of apprehending the danger. We lack a mechanism to link our ever-increasing demand to the worsening environmental crisis, to assess what the cost of growth will be. At the same time our foreign policies continue, directly and indirectly, to spread misery overseas, adding to the burden of desperate refugees who will seek our shores in the future.

The fundamental question of sustainability is at the heart of the matter. What do we mean by sustainability? It is the overriding desideratum that we should live without destroying future prosperity. Sustainability is integral to our sense of ourselves as a nation because we must achieve it if we are truly to call this continent "ours".

If we are ever to break the pernicious connection between our environmental mismanagement, disgraceful treatment of refugees and other victims, and lack of direction when it comes to the question of how many people we can support, we need to face some hard truths. If things do not change, it will become more and more difficult to sustain even the number of people we now have, let alone take in the vast number of refugees our foreign policy threatens to create in the days ahead. Right now we pin our hopes on the delusion of Fortress Australia – a mirage-like refuge where negligent people can hide their heads in desert sands. What we need, I believe, is to face the future with all the toughness of mind and humane care we can muster, and practise a humanism that recognises the cost of the choices we make as well as the moral necessity of trying to create the greatest possible good.

In essence, this essay tries to expose the heartfelt falsehoods that keep us from seeing the truth of our situation. These lies are close to the bone, they go deep in the Australian legend. They are the things that lead to slanging matches at family get-togethers and flying fists in pubs. So let us begin resolutely by examining the greatest lie of them all.

No lies are as potent as the lies we tell about land and people. Because they justify or deny the access of individuals – and indeed of entire societies – to acres and resources, they often take on a legal status. Australia's founding lie was as wide and all-encompassing as the continent itself: *terra nullius*, the myth of the empty land, whose Aboriginal inhabitants of some 47,000 years tenure had, under British law, no rights to their country whatever. The land was literally taken over in *toto*, and Australia has been a colonial country, in the truest sense of the word, ever since. For the continent was colonised by an immigrant people whose food and mores, outlook, laws and social organisation originated elsewhere. To a great extent that situation still prevails, and it has ramified in our history into a thousand misunderstandings and errors. It lies at the heart of most of our environmental woes, and it is central to the dilemma we face in defining ourselves as a people.

In a sense, colonisation resulted in a classic mismatch – Australia is the ultimate round peg in a square hole; the round peg being the colonial insertion of a European people, their domesticated species and their laws, while the square hole was the environment of the continent. Despite our long-held beliefs to the contrary – with all of our aspirations to create European gardens, drought-proof the continent and create an Anglo-Saxon society at the end of the world – it is the "square hole" of Australian nature that remains immutable. After two centuries of being bashed up by drought, flood, fire and ecological catastrophe of every kind, we are slowly discovering that if a fatal mismatch is not to occur, it is the shape of the round peg – the colonial insert – that will have to change.

What Australians have needed are people at the wheel who can change society's direction and turn it towards an genuinely emerging post-colonial awareness, a condition that allows us to strive for an adaptation of law and other cultural baggage to Australian conditions so that we can finally end the colonial period of our existence. A native Australian, Koiki (Eddie) Mabo, accomplished more than anyone in the last 200 years in achieving just that when, at a single stroke, he exposed to the light of scrutiny Australia's founding lie of *terra nullius*.

Undoing all the evil the lie of *terra nullius* has engendered, however, has been a slow and sometimes uncertain business. The very land-claims process that Mabo initiated has left many Aboriginal people disillusioned and frustrated. Indeed, progress has been so poor in this area that Rick Farley argued in his 2003 Australia Day Address that the entire land claims process had been captured by wealthy whites; and certainly the lawyers have done better out of it than anyone. But even if this charge is true, it should not detract from the magnificence of Eddie Mabo's achievement; for ever since he won his fight, in court cases and in subtle shifts of public attitude, our founding lie has started to be drained of its power. In a sense, the decolonisation of Australia's legal system began in 1992 with Mabo, and one day, as a result of his actions, a better, more equitable and less colonial Australia will exist. Perhaps a century from now Mabo Day will mean more to our nation than any anniversary of Federation.

At its founding, European Australia was one vast prison, and prisons are almost invariably institutions founded upon falsehood. Not only are they filled with dishonest individuals, but those who build and pay for them are liars too. Otherwise honest citizens tell themselves that jails are a necessary evil that serve to reform criminals and to keep society safe; yet society rarely gets safer as jails proliferate, nor do rates of recidivism decline. Indeed, most often the reverse occurs and crime rates spiral as the number of penitentiaries increases.

Yet European Australia was a far from typical prison, because it was the only prison in the history of mankind that I'm aware of which actually fulfilled the hopes of those who built and paid for it. A one-way journey to "Botany Bay" often turned criminals into respectable farmers, merchants and politicians; and by emptying the hulks and prisons of Britain, it kept the slow burn of social unrest from reaching the flashpoint of revolution which would have turned the islands into a bloody battle-ground. It was a prison that actually benefited many of those who entered it. But it was also an ugly, brutal place.

Young Charles Darwin saw all this very clearly during his sole visit to Australia in 1836. He toured penal institutions in New South Wales and Tasmania and heartily abhorred what he discovered. He was shocked to find that "children learn the vilest expressions … if not equally vile ideas" from the convicts, and felt the horror of being waited upon by someone "who the day before, perhaps, had been flogged, from your representation, for some trifling misdemeanour". Indeed, he could find no respite from the all-pervasive brutality and degradation of the thief colonies. Even as he travelled into the Blue Mountains to study their geology, the clanking of the chain gangs rang in his ears, and when he visited the genteel parlours of Hobart he was told of the destruction of the Tasmanian Aborigines as the convicts and their masters stole black land. Yet for all the horror of convict Australia, Darwin also knew that his life as a country gent in Britain could not be maintained without this terrible place; for it was the deportation of criminals to Australia that made his bucolic bliss possible. It was perhaps this sense of being personally dependent on the perpetuation of a terrible wrong that made Australia so repugnant to him. He left "without sorrow or regret", his mind too filled with revulsion and homesickness for him to have learned anything from such natural wonders as the platypus and the kangaroo.

Unlike Darwin, many European Australians still do not understand the implications of continuing to live in a colonial society. The Man from Snowy River is an archetypal Australian hero – one of the brave Aussies who tamed the rugged land. He sits side by side with the archetypal stockman in our constellation of national icons. We even have an entire museum at Longreach, Queensland devoted to the adulation of such men. Yet our worship of the self-reliant stockman neatly sidesteps the fact that the men of the cattle frontier were the shock troops in our Aboriginal wars. As Henry Reynolds has so amply demonstrated, during our frontier wars – the only wars ever fought on Australian soil – thousands of men, women and children were killed in battle or murdered in cold blood.

There is a deep current in our colonial Australian society that resists these simple facts and clings to the great founding lie. Although the scientific and historic proofs are numerous and incontrovertible, the palpable falsity of *terra nullius* has not been overturned by appeal to reason; instead, opponents have often turned to the denigration of indigenous culture as a way of distracting themselves from modern Australia's colonial status. Through such denigration, they hope to engender the feeling that the Aboriginal people are inferior and unworthy – perhaps not even fully human. There is no better recent example of this than the claim in Pauline Hanson's book *One Nation* that many Australian Aborigines were cannibals. The book cites as evidence sensationalist, popular novels written during the late nineteenth century and set on the Queensland frontier – works that have no anthropological credibility whatever and which portray bloodthirsty natives attacking Europeans and Chinese and eating their flesh with relish.

Hanson's claim was itself attacked violently, her accusation of cannibalism emphatically dismissed by many academics as evidence of Hanson's contemptible ignorance. And yet the debate troubled me deeply, for I knew that even a casual perusal of the Australian anthropological and historical literature indicated that cannibalism was indeed practised in some Aboriginal societies, albeit in a very different way from that claimed by Hanson. Perhaps the most unimpeachable example is found in William Buckley's account of his thirty-two years living with Aboriginal people before European settlement in the Geelong area. Buckley identified two kinds of cannibalism, both of which he seems to have witnessed. The first kind appears to have been geographically widespread in Aboriginal Australia and involved the consumption of portions of a dearly loved relative (particularly a child) as part of the funeral rites. This form of cannibalism is an act of the deepest love and affection. It perhaps finds its closest parallel in Western society in the Catholic understanding of the rite of Communion. The second kind of cannibalism took the form of ingesting the parts of warriors killed in battle in order to obtain their bravery and strength. This seems to have been less widespread even within a single clan, for Buckley claims that this form of cannibalism revolted many of his Aboriginal acquaintances, who instead rubbed the fat of slain warriors on their skin to achieve the same end.

Anyone knowing of these practices at the time of the One Nation furore was faced with a dilemma. Should they lie – denying Aboriginal cannibalism – and so condemn Hanson's support for Australia's founding lie; or should they confirm the apparent historical reality of Aboriginal cannibalism, yet still insist that the founding lie must be overturned? While the debate was raging there was no room for equivocation, and every academic I saw interviewed over the issue chose to deny the evidence of cannibalism in Aboriginal societies, which most must surely have known of. There was, however, one notable exception – the remarkable Les Hiatt, a respected senior anthropologist. His thoughts, published in a letter to the *Sydney Morning Herald*, deserve to be reproduced at length here:

> The debate on Aboriginal cannibalism generated by the Pauline Hanson book has proceeded on the assumption by both sides that the eating of human flesh is shameful. Denials and affirmations of its occurrence are accordingly seen as attempts either to elevate or to lower the image of Aboriginal people in the estimation of white Australians.
>
> The assumption of shamefulness needs to be challenged. In many parts of Australia, Aborigines recovered the bones of recently buried relatives and kept them until the pain of bereavement abated. Sometimes, as in the case of small children, they refused to part with the bodies. In rare circumstances, as when young warriors or women fell in the course of battle, their anguished kin first attacked their own bodies and then ate the flesh of the deceased.

An eyewitness account of such an event was given last century by a Victorian assistant protector named C.W. Sievewright and was published in the *Victorian Historical Magazine*, 1928, pp. 168–170.

As Kenneth Maddock has indicated (*Herald*, 23 April), necrophagy in Aboriginal Australia is well attested. Those who consumed the bodies of their loved ones did so while they themselves were consumed by grief.

Such ancestors are not necessarily more shameful than ours, who taught us to bite our own lips and choke on our tears.

The feeling of those times – the time of One Nation – was very like being in a war. It was a winner-take-all battle between a reactionary old Australia with its belief in *terra nullius* and an emerging, post-colonial and reconciled Australia. And as in any war, the first casualty was truth, and it was those on the liberal left as well as those on the right who were, almost wholesale, willing to sacrifice that truth for contingency. It's a pattern I see repeated over and over again in our history.

What made matters worse in the Hanson fracas was that few questioned the fundamental assumption behind the debate. Was it really right or just that the descendants of people who may have practised cannibalism be shorn forever of all rights to their lands and resources? Does the act of cannibalism make a person less than fully human? Instead, at the height of the Hanson debate, lie piled upon lie, and bitter hatreds and divisions were spawned among Australians.

This appalling situation was made worse by the fact that generations of Australians had turned their backs on their own country's history. They might have secretly feared that they would discover in the annals of the Australian pioneers evidence that their nation was born in a bloodbath of inter-racial violence – the frontier wars – and perhaps even that their own families had been involved. This was not a lie, but it was a monumental refusal to look truth in the face and to learn where we had come from.

While such fears may not be unfounded, to ignore our history is to live in darkness. For as well as discovering appalling truths in our past, we find the most extraordinary examples of deep humanity. Indeed, I think that in parts of our history we can discern a clear template for creating a better Australia. This is particularly true of our founding texts – the five principal narratives by Phillip, Hunter, White, Collins and Tench dealing with our earliest history, the half-decade from 1788 to 1793. The accounts are as variegated as they are revelatory, and to read them is to discover the unexpected; for the story of Australia's founding embodies a vital, redemptive thread in our history, which seemed lost for so long but which now, in the wake of Mabo, we are beginning to see the significance of. And, if we are talking about lies versus truths worth living by, I also think that the first four years of European settlement in Australia represent a unique experiment in discerning the human condition.

The terms of the experiment were dictated by the elements that had been thrown into the mix. On one side were over a thousand colonists from the far north of half a world away. They included 736 convicts, the majority drawn from the slums of London, and then, overseeing them, a group of Royal Marines under the command of Governor Arthur Phillip. The contrast between the two groups of whites could scarcely have been greater: the convicts were for the most part starving, illiterate and lost, while in Phillip and the marines we have some of the highly sensitive minds that come out of the culture of the European Enlightenment. Indeed, in individuals like Lieutenants William Dawes and Watkin Tench we find spirits whose humane attitudes and whose understanding of the nature of Australia were not to be met with again in the antipodes for two centuries.

Arthur Phillip, Australia's first governor, was the son of a German migrant from Frankfurt who might possibly (although this now seems unlikely) have had Jewish ancestry. He was a fluent speaker of Portuguese, having fought with them against the Spanish as a mercenary. As this history may suggest, he was also someone with a keen sense of natural justice and an appreciation of cultural difference. By ordering that a portion of every catch of fish the Europeans made should be given to the Aborigines, he recognised the primacy of the Aboriginal claim to resources. Phillip was also ageing (he was forty-eight when the fleet sailed) and in poor health, and by 1791 his continued physical frailty was beginning to show in some of his decisions about how to deal with the Eora people of the Sydney Harbour area.

Lieutenant William Dawes has been described by an eminent Australian historian as a "man of letters and a man of science, explorer, mapmaker, student of language, of anthropology, of astronomy, of botany, of surveying, and of engineering, teacher and philanthropist". He was the colony's official stargazer and was placed in charge of its ordnance. His astronomical observatory had to be located outside the settlement so that the lights of the campfires did not disturb his observations, and he hit upon a point to the west of the settlement which today supports the southern pylon of the Sydney Harbour Bridge and is known as Dawes Point. This advantageous geographical location – somewhat removed from the main settlement – gave Aboriginal people the opportunity to visit a European without having to enter the main camp, and Dawes began to build a unique relationship with the Eora. His diary, which is held in the Institute for African and Asian Studies, London, documents his attempt to learn their language, and at the same time reveals the development, Professor Higgins-style, of an intimate bond with a young Aboriginal woman called Patyegarang. "*Matigarbárgun náigaba* – We shall sleep separate"; "*Metcoarsmady minga* – You winked at me." The incoherent strings of words in parallel languages run on, building a bridge on which black and white could have built a greater mutual understanding.

A signal event assured, however, that this was not to be. In December 1790 a young Aboriginal man speared the colony's "game-keeper", McIntyre. ("Game keeper" was one of those terms whose meaning had become inverted when it was transposed to this new colony in the antipodes. The Aborigines saw McIntyre for what he was, a poacher of their wildlife.) Phillip, in an uncharacteristic outburst brought on by ill-health and exhaustion, ordered Dawes and Tench to bring six Aborigines to him in retaliation or, "if that should be found impracticable, to put that number to death".

Dawes reluctantly joined the first party sent out to achieve this horrible retribution. But when the venture failed and a second attempt was ordered, he had to make a choice – in effect a choice between natural justice and obedience to a superior. He fully understood that disobedience would be seen as an act of treason that could see him swinging from a gallows without delay; yet he chose the side of Patyegarang and justice, making it widely known that he would refuse to join any such future punitive expedition.

Governor Phillip stayed his hand against his lieutenant until 1791 when the *Gorgon* arrived to relieve the marines and Phillip's command. Then, despite Dawes' desire to remain in Australia, he was ordered "home" to England, and he and Patyegarang were separated forever. If Lieutenant William Dawes had been allowed to join our first settlers, what a contribution he might have made! Still, his life in the wider world was not wasted. He went on to become a tireless anti-slavery campaigner and promoter of full rights for coloured people.

Dawes' greatest friend in the new settlement was without any doubt the extraordinary Watkin Tench, our nation's first chronicler and the author of two highly readable and compelling accounts of Australia's first four years. He was widely acknowledged as "the most cultured mind in the colony", and such was his enlightened and humane outlook that, despite the prejudices of his times, by the end of his sojourn among the Aborigines he could profess that, "Man is the same in Pall Mall as in the wilderness of New South Wales."

Tench was astonished by the fearlessness of Eora leaders like Bennelong in the face of overwhelming odds. He was also deeply affectionate to many of his other Aboriginal friends. Indeed, while his view of indigenous people was unhindered by any concept of the "noble savage", he held some

Aborigines in deep regard – not because they were black or different, but simply because they were remarkable people. After leaving Australia in 1792 he would go on to fight in the French Revolutionary wars, and was serving on the *Alexander* under Captain Bligh when he was captured and sent into rural confinement in Brittany. At that time, to keep a diary as a prisoner of war opened one to charges of espionage and thus death. Yet Tench took the risk, and his *Letters from France* were published as his third and final literary production. It was the only one of his published works not dealing with Australia. But the links were still there, for Tench tells us that his secret diary was written in part in "the language of New Holland" so that its contents could not be understood if found by his captors.

On the other side of this remarkable human experiment that marks Australia's founding were the Eora. Because they left almost no written accounts, we know less about them as individuals. Yet among them are statesmen like Colbee and warriors such as Bennelong and the resistance leader Pemulwuy ("Man of the Earth") who for years fought the Europeans. Bennelong, by virtue of his quick wit and bravery, enchanted Tench. "Love and war seemed his favourite pursuits," wrote the marine about his Aboriginal friend, whose "powers of mind were far above mediocrity". On the fateful day when Governor Phillip was speared at Manly Cove, Bennelong (who had earlier escaped from the settlement after being kidnapped) inquired by name for every person whom he could recollect at Sydney; and among others for a French cook, one of the Governor's servants, whom he had constantly made the butt of his ridicule by mimicking his voice, gait and other peculiarities, all of which he again rehearsed with his accustomed exactness and drollery.

Three important lessons shine through to me from this nascent Australia. The first is a confirmation of the common humanity of all people – as strong a confirmation as you could wish for of the absolute necessity of living by a humanist or at any rate a humane creed. The reason I say this is because the two peoples who met on that momentous day in 1788 – the Aborigines and the Europeans – had been separated from each other for longer than any other human cultures on our planet. For 60,000 years – perhaps half the span of our species' tenure on earth – they had been cut off from each other, living on isolated and very different landmasses at opposite ends of the globe. They had developed separate languages and cultures, different skin colours, gene frequencies and facial features. But despite it all, recognition and understanding were immediate, for so strong is our common bond that 60,000 years of separation melted away in a moment. A smile was a smile. An uncertain glance, an act of friendship, a shout of hostility or fear, a sexual overture – all were instantly comprehended.

This was true from the first instant of contact – and here it is, related to us by the remarkable Watkin Tench, who recorded exactly what happened when that 60,000-year separation was finally ended:

> I went with a party to the south side of the Harbour [Botany Bay] and had scarcely landed five minutes when we were met by a dozen Indians, naked as the moment of their birth ... I had at this time a little boy, not more than seven years of age, in my hand. The child seemed to attract their attention very much ... and as he was not frightened I advanced with him ... at the same time baring his bosom and showing the whiteness of the skin. On the clothes being removed they gave a loud exclamation and ... an old man with a long beard, hideously ugly, came close to us ... The Indian, with great gentleness, laid his hand on the child's hat and afterwards felt his clothes, muttering to himself all the while. I found it necessary, however, by this time to send away the child, as such a close connection rather alarmed him, and in this ... I gave no offence to the old gentleman. Indeed it was but putting ourselves on a par with them, as I had observed from the first that some youths of their own ... were kept back by the grown people.

The second great lesson I find in those first four years is that they gave reason to hope that black and white could co-exist – and indeed forge a new and distinctive nation in Australia. A personal friendship grew up between Governor Phillip and Bennelong who, with his clan, was given ownership of one of the first brick buildings erected in the settlement. Catches of fish were shared and, as Watkin Tench witnessed, a feeling of mutual respect grew between the European intellectuals and the Eora.

What killed this earliest of reconciliations was a struggle over resources. Hitherto the Europeans had lived out of government stores – on years-old pickled pork and weevily flour. But then, in February 1792, Governor Phillip made the first grant of land in the new settlement, giving thirty acres near Parramatta to James Ruse. He must have known that it was not his land to give away, for the colony's judge advocate, David Collins, had found clear evidence of individual ownership of particular pieces of land among the Eora, and of such properties being inherited down the generations.

As the Europeans started to appropriate land for their own use as farms, in creating towns and shaping gardens, for the first time the question of who owned the land – on the broad scale – became an issue. Angry Aborigines who met Europeans on the Parramatta road demanded to know what they were doing. "This is our country," was the continual refrain. The government's only response was to increase the military presence in the newly cultivated areas. When an Aboriginal boy was killed by the settlers, the perpetrators were brought before Judge Collins; yet he released them because he was uncertain about the status of Aboriginal people and their testimony under English law in the colony. Justice Aboriginal-style followed, and then the bloody battle was on.

The third lesson I take from Australia's founding is that a deep humanism as well as enlightened attitudes can survive – indeed flourish – in brutal and divided times, in a distant colony where the Europeans faced starvation in a prison of their own making. The learning, compassion and sheer intellect of the best of Australia's early European settlers are still astonishing. They possessed a world view rooted in the teachings of figures like Adam Smith and David Hume. Individuals such as Tench, Dawes and Phillip shared a humanism and speculative breadth of mind that admitted no prejudices and few preconceptions. As we read Tench on the Eora we forget, after a while, that we are reading of "naked, wild savages" and instead we see these people first and foremost individually – we see the faces and habits of Coleby, Bennelong or Araboo, all dealing with their intensely human dilemmas in brave and intelligent ways. The humanist perspective could not instantly right the wrongs of history and circumstance, but its survival, and the good it did, shines out to us today as we grapple with problems that seem insoluble.

Among later, often Australian-born explorers we find no such sympathy. Frontiersmen such as John Forrest and David Carnegie, who were charting the Australian inland generations after Tench, could see in the Aborigines nothing but impediments; their humanity had receded completely. It was this generation, growing up without the benefit of an Enlightenment education, which took the land by force and squeezed Australia into the mould we know today.

Tragically, it was in this turbulent period that the new pattern of interaction was set – dispossession of a proud people, then resistance, followed by reprisal. It has taken us 200 years to begin to get back to anywhere near where we started – to a mutual respect and a willingness to talk and listen. And the talk and negotiation today is over the same issue as it was then – land and resources. *Terra nullius.*

So at the beginning there was that rarest of things. An "honest" prison, gaolers who were a testament to the high values of their age, and a real hope that black and white might make a go of it together in Australia. No armband history, black or white, just reason for hope, and acknowledgement that there was no other possible way of moving forward. For in those infant days at Sydney Cove, neither black nor white wielded absolute power. Modern Australia was thus of necessity born through negotiation and reconciliation.

While these human dramas were playing themselves out in the colony, fundamental changes were occurring in the Australian environment which unfortunately attracted very little attention. Yet it is vital that we examine them here because there has been a lot of unacceptably woolly thinking about the environmental crisis in Australia. If we are to take Australia's environmental crisis seriously and, as I will suggest here, fight it as if we are fighting a war, we must base our action plan on the best science we have rather than prejudices or assertions. This issue is particularly important because the woolliness extends to the heads of some of those who lead the Australian environmental movement.

There is a man in South Australia by the name of Dr John Walmsley who is deeply committed to the environment and who is often seen parading about with the skin of a feral cat on his head. He does this because he believes that cats are a terrible scourge on the Australian landscape, and in this belief he is joined by a good many professors in universities and curators in museums. One museum curator has reported, in condemning the cat as a major menace, that felines kill billions of native animals each year. It's a frightening figure that has led many to see cats as major contributors to Australia's biodiversity crisis. Yet is the moggy really responsible for the extinction of Australian natives? Clearly cats are efficient predators, and anyone wishing to see large numbers of native birds and animals in their garden would be well advised not to make their home the centre of a feline's territory. Yet beyond their hunting finesse, there is little evidence that cats have exterminated any species in the Australian environment.

The role of cats takes pride of place in this discussion of environmental lies because they were probably the first introduced creatures to spread widely after being brought to Australia by the Europeans. Cats arrived with the First Fleet and many soon went feral. Although we have only an imprecise idea of their rate of spread, it must have been rapid, because explorers who questioned Aborigines in central Australia about cats in the late nineteenth century were told that the creatures had been present in the Australian desert for a very long time – longer than anyone could remember. This accords with the well-documented rate of spread of other exotics such as foxes and rabbits, which are similar-sized creatures and thus about as mobile. Both species took only a few decades to reach the zenith of their distribution on the continent, which extended from Sydney to Perth, and north of Alice Springs to Melbourne (though favourable seasons allow occasional extensions north). Given this, it seems reasonable to believe that cats had spread throughout the continent by the 1840s, some sixty years after their introduction.

So which native species were in decline in eastern Australia by the 1840s? Curiously, very few of our native creatures destined for extinction show a pattern of widespread decline at this early date. The only likely candidate is the white-footed rabbit-rat (*Conilurus albipes*), a native rodent the size of a squirrel which I will have more to say about in a moment. Almost all the other well-documented, now-extinct species (at least on the mainland), such as the eastern quoll, lesser bilby and Tasmanian bettong, overlapped with cats for at least fifty years, and in many cases for over a century. As extinctions, or at least drastic declines in animal populations, usually occur soon after the causative factor is introduced, this suggests that cats were not, by themselves, the primary agents of extermination.

There is one further very important proof in the argument that cats have not caused widespread elimination of species in Australia. Tasmania and Kangaroo Island are two of Australia's largest offshore islands. They are varied and large landmasses, comprising a diversity of environments that support rich marsupial, bird and reptile faunas. Both also have feral cats, yet with the exception of the thylacine and dwarf varieties of emu (which were hunted into extinction by human beings) neither has suffered a single extinction. Instead, cats cohabit with the full faunal diversity of both islands.

It's true that on far smaller islands cats may have a greater impact. The 5,000-hectare Faure Island in Shark Bay is dry and flat. Before the introduction of cats, goats and sheep, it had barred bandicoots, Shark Bay mice and woylies (a kind of kangaroo-rat). Cats may have played a role in the extinction of such isolated and vulnerable populations of native mammals, especially after the island had been de-vegetated by goats and sheep, eliminating places to hide. Cats can also frustrate efforts at reintroducing native species, for the captive-bred marsupials have no experience at all of complex natural environments or predators, and are sitting ducks for any half-competent predator.

How to reconcile this information with the dogmatic belief of some environmentalists that cats are a major − if not *the* major − threat to Australian wildlife? Seeing a cat with a native bird in its mouth may lead a member of the public to the erroneous conclusion that cats cause extinctions in Australia, but such thinking would seem naive in a university professor. My sense is that the nub of the matter hangs on prejudice. Cats are the Arabs of the animal world. The world is divided between cat-lovers and cat-haters, and perhaps the majority of those who assert that cats have caused extinctions in Australia are simply cat-haters who have allowed their prejudice to override their scientific reason.

If the extinction of large segments of Australia's fauna cannot be slated home to cats, a multitude of other causes set in motion in First Fleet times are likely to have played a part. In the early days, two, I believe, were particularly important: sheep and *terra nullius*. In journal after journal kept by Australia's pioneers you can find records of the decline of the native mammal fauna as a consequence of sheep grazing. In South Australia, for example, gross overstocking of the Flinders Ranges and regions south, from around 1860 to 1880, closely correlates with the extinction of medium-sized native mammals. In Victoria the wholesale importation of sheep into Australia Felix, as Victoria's volcanic plains were known, led to massive environmental change. Before sheep, the most abundant native animal on the volcanic plains was a large native mouse whose bones have been found by the thousand in caves of the region. Yet it became extinct so rapidly after sheep arrived that it was never collected or recorded as a living animal by a scientist or settler, and to this day this once conspicuous creature lacks a scientific name. At the confluence of the Murray-Darling it was a similar story. There, Gerard Krefft, director of the Australian Museum, collected large numbers of mammals in 1857–8, at the same time recording the wave of extinctions that occurred as the sheep frontier advanced from the south.

It is important to realise, however, that what people were recording here were mostly local eliminations. Where there were no sheep, the native species persisted, and not all of them were adversely affected. Other, more widespread changes would deliver the *coup d'état* to Australia's marsupials and native rodents.

Among the most important of the continent-wide changes with their roots in First Fleet times was the cessation of Aboriginal burning of the land. Fire was by far the most important tool possessed by Aboriginal people to manage their land, but with the concept of *terra nullius* firmly embedded in their minds, the pyrophobic Europeans took control of the fire-stick and the vegetation changed. From the first the Europeans muttered about the damage that could result if Aboriginal fire was used as a weapon, and laws were passed prohibiting the Aborigines from burning the land. There are few detailed historical studies of the impact of this change, but I have suggested elsewhere that it is the major cause of mammal extinctions in central Australia. The aftermath of the January 2003 fires has brought this point home in the most tragic manner, for the futures of two of Australia's most endangered creatures – the corroboree frog and the mountain pygmy possum – hang in the balance after their populations were ravaged by the blaze. Indeed, some wildlife workers fear that the corroboree frog was made extinct by the fires.

The fate of one species in particular, however, gives us a clue to the general pattern of decline that may have followed the end of native fire. When the First Fleeters established the settlement at Parramatta, they found that their stores were being raided by a large, handsome native rat. Gnar-ruck, the Eora called it, and today we know it as the white-footed rabbit-rat. The strange thing about the Gnar-ruck is that just about everywhere there are records of it, the first European record is also the last. So it is that the drawing made of it by an anonymous First Fleet artist around 1791 is both its first and last record from Sydney. It was seen near Melbourne in the 1830s and in the Adelaide area in the 1840s, but after that it vanishes. Before European contact it was to be found over a vast area of eastern Australia. It preferred grassy woodlands and made its nest in hollow logs, which it packed full of dry grass. Such woodlands were burned regularly by the Aborigines, and the rat seems to have flourished in this maintained environment. As soon as the Europeans moved in and altered the fire regime, the rat vanished. It is as clear a case as any of an Australian mammal whose extinction follows closely the march of the doctrine of *terra nullius* across the land, and the fire change that followed in its wake.

The greatest animal scourges to assail the Australian environment – the rabbit and the fox – post-date the first fifty years of settlement, so we will return to discussion of their impact presently. For now I would like to examine the nature of European society as it manifested itself in Australia in the fifty years following first settlement. Its development suggests intriguing insights into what became the Australian attitude to cultural diversity. It's worth reminding ourselves that until the 1830s, not only could you meet naked and armed Aborigines in the streets of Sydney and other Australian towns, but you could also be ferried across Sydney Harbour by a West Indian man and his European wife or their son. You could buy your tea from an Anglo-Chinese currency lad, and in those wondrous times you could even enjoy a Muslim festival in the streets of the harbourside city.

To immigrants arriving in Australia after the Second World War, it might have seemed that they had come to a dull, stodgy clone of Britain located at the ends of the earth. At least that's the impression that floats wordlessly about us as we look and marvel at the cultural transformation that has occurred in Australia since the 1960s. I clearly remember the first pizza shop to open in my part of Melbourne, the first Mexican restaurant and the first avocado to appear in the local greengrocer's. Six o'clock closing, mutton and potatoes, chops and three veg. were the order of the day during my childhood. In the year I was born – 1956 – Tasmania's Wrest Point Hotel opened. That icon of Australian eating offered on its inaugural menu just three choices for mains – lamb, beef or pork, with strawberry ice cream for dessert! Culinary exoticism was provided in my childhood by a Chinese restaurant run by a diminutive woman from Clapham, England who, if we wanted take-away, ladled chow mein and rice into the pots and pans we brought in. The story we love telling each other is that somehow a monotonous, derivative Australia was rescued from itself by an influx of colourful and tractable migrants from around the world. The only problem is that it is all a terrible lie. Australia was multicultural before ever it was insular, and all of that diverse food – it would have come anyway.

Before we visit the roots of Australian multiculturalism in early nineteenth-century Australia, it's worth diverging briefly to examine this most delicious lie of all – that by bringing a global cuisine to Australia, immigration rescued us from cultural death by British stodge. When the benefits of multiculturalism are discussed, food is usually near the top of the list. The unadulterated meat and three veg. of 1960s Australia was banished, the argument goes, by a flood of exotic and delectable cuisines. But the reality is that food became internationalised in the closing decades of the twentieth century. Anyone who has travelled to the culturally homogeneous regions of Europe, or even in parts of

China, will know that the most exotic of restaurants can now be found in the most unlikely of places. Even in such a remote locale as New Zealand, which has experienced little migration, you can find restaurants serving the most varied ethnic foods. Certainly a critical mass of discerning migrants may have contributed to the excellence of contemporary Australian cuisine, but, difficult as it is to swallow, I believe that globalisation would have changed Australia's eating habits regardless of the level of migration, or where the migrants were drawn from.

Even so, there is no doubt that Australia's immigration policy has had a profound social impact on the nation over the last fifty years, for Australia now has one of the largest immigrant populations, relative to its overall population size, on the planet. Nearly 25 per cent of Australians are foreign-born, as opposed to 19.2 per cent in Canada, 11.7 per cent in the United States and 3.4 per cent in the Netherlands. If we consider both first- and second-generation Australians, the number rises to 44 per cent. Not since the gold rush has the proportion of immigrants in Australia's population been so high.

The eternal tensions between the latest wave of migrants and those already settled here are of great concern to social commentators and to shock-jocks such as Stan Zemanek. Yet ever since the First Fleet convicts stepped ashore, there has been friction between new and old arrivals. And as we have seen, the essential oneness of humanity that was confirmed so remarkably in First Fleet times has ensured that the Wogs, Balts and Slopes of yesterday are the mates and "dinkum Aussies" of tomorrow.

Yet somehow the argument is still being put that the latest group of immigrants is different. Thoughts like "the Chinese never mix" or "Muslims will always be different" are whispered or secretly pondered countless times each day around the nation. Past experience seems to make no dent on these views; it is as if they are products of the reptilian brain, these knee-jerks of prejudice. Such nonsensical, such abysmally pessimistic and ungenerous manifestations of human blindness are deeply depressing, all the more so because history has already proved them so abundantly wrong. For, as we will see, the Muslims and the Chinese have been part of the Australian cultural mix almost from the beginning.

The best lies always aspire to the condition of minimal truth, and there is a shred of verisimilitude to the myth of an "Anglo" foundation of Australia, for the majority of First Fleet convicts were certainly British, or Londoners to be more precise. In fact, around 440 of the 736 hailed from the great British metropolis. Nonetheless, by the late eighteenth century London was one of the world's pre-eminent entrepots: its dockside areas teemed with people from foreign lands, many of whom were caught up by the long arm of the British rule of law. Who knows what was the "true" racial identity of many of those early convicts? But even in the First Fleet we don't need to look far to find diversity. Of course, there were the ubiquitous Irish and Scots with their complex degree of separation from the "Anglo essence", but among the arrivals must also be counted Jews, West Indian Negroes (2 per cent of the total), a French chef and an American sailor, Jacob Nagle, who crewed Governor Phillip's row-boat. The Second Fleet, arriving in 1790, brought further diversity — including more Negroes and Jews, along with Canadians and Germans and a possible range of other nationalities.

As early as 1793 Muslims formed part of the mix to be seen in the streets of Sydney, for Lascar (Asian) sailors from the merchant vessel *Shah Hormuzear* are found in the colonial records of this time. Islamic sailors were even more visible in 1806, when Shiite members of the crew of the *Sydney* celebrated the festival of Husain. They laboured for over a fortnight constructing a magnificent replica of Husain's coffin, which was the centrepiece for varied performances that went on for a week. The town watched on respectfully, the populace awed by the munificence and splendour of the occasion. Indeed, it seems to have been one of the most spectacular "events" Sydney had seen since its founding.

And this apparition of a non-British world also had a human face. One of the most colourful identities of early Sydney was Billy Blue, a Jamaican Negro who married a European woman and ran the harbour's first ferry service, a tradition carried on by his son. Billy arrived as a convict in 1801, but the following year was appointed water bailiff. He was famous for his "jive" which, in the best Australian tradition, took the high and mighty down a notch or two.

The popular and newspaper accounts of Sydney at this time are striking for their lack of racial slurs or any of the other stigmata of racial tension. Far more typical, indeed, are descriptions like this one by Peter Cunningham, who could say of his home town:

> French, Spaniards, Italians, Germans, Americans … all add to the variety of language current among us … [there are] several ingenious and industrious [Chinese] individuals …[who] flourish as members of our community. Over a snug cottage at Parramatta, for example, may be seen the sign of "John Shan … carpenter, and dealer in groceries, teas etc." with his tidy English wife and a group of Anglo-Chinese descendants around him.

A walk along the waterfront reinforced the point:

> As you stroll along the picturesque shores of our harbour, you may often be melted with the wild melody of an Otaheitean love song from one ship, and have your blood frozen by the horrific whoop of the New Zealand war dance from another.

So wrote the enchanted Cunningham. Sydney in the 1820s had the feeling of a place where almost everyone is a new immigrant. It was the crossroads of the South Pacific, and it seemed to revel in the mix of peoples that composed its population.

The conservative, overwhelmingly "Anglo" Australia that would greet first the Chinese, then after World War II the "Balts, Wops and Dagos", so dismally seems a long way from this accepting, variegated Australia of the 1820s. The question as to what wrought the transformation is a profound one, for it is germane to understanding Australia's great social dilemma today.

A significant part of the answer can be found in the mid-nineteenth-century pattern of immigration, which was dominated by immigrants from the British Isles and led to the founding of cities such as Melbourne, Adelaide and Perth. The gold rush, of course, brought a greater diversity, but much of this cultural variety seems to have been absorbed back into the mainstream by the time of Federation, when Australia was exhibiting remarkable uniformity. And, in an age when communications over large distances were difficult and belated, it cannot be overestimated how such uniformity meant that the mainstream was becoming peculiarly narrow.

Australia's position in the antipodes meant that throughout the nineteenth century British colonial education was not what it was in Edinburgh or Oxford. Most currency lads and lasses received only rudimentary schooling, and this led to a much more insular outlook than that of the intelligentsia of the First Fleet or even that of the upwardly mobile "emancipist" convicts. The explorer John Forrest, later Premier of Western Australia, was, alas, a representative figure. Not only did he view the Aborigines as unworthy of elementary respect or consideration, but he was ignorant of the very European culture that served as the basis for his sense of superiority. When he made the journey from the west to Melbourne towards the end of the nineteenth century, it was his first glimpse of a real city. He was singularly unimpressed by the cable trams, the cathedrals and bookshops, but he did trudge in his explorer's boots up Collins Street to see the one thing that meant something to him – the statue of those doomed adventurers Burke and Wills. Countless other Australians would have done likewise, for Forrest was the product of life in a frontier society whose values and social mores were a provincial version of Victorian dourness and high seriousness, stripped of the Enlightenment breadth of mind that had preceded them.

Worse still, by the second half of the nineteenth century, the Western intellectual milieu had itself changed. When Watkin Tench came to Australia, it seems that he carried a copy of Ferguson's *Essay on the History of Civil Society*, which had been published in 1767. If he did not, he certainly knew it intimately enough to quote from it. Such a book would have been all but unknown in the Australia of the 1860s, and a century after its publication the attitude of civilised tolerance it encourages had become equally remote from Australian consideration. The marvellous works of Ferguson, Hume and Smith, which Manning Clark could feel the afterglow of with the First Fleet, had receded. Even a Victorian utilitarian and liberal like John Stuart Mill rapidly lost his first gloss in Australia, and a new form of "science" – social Darwinism – usurped the place of such wisdom in shaping society's attitudes. We must never forget that the cultural milieu that gave rise to Australia's Federation was not the enlightenment of Hume or even the liberalism of Mill, but the Social Darwinism of the phrenologists and other racists (and this remains true whatever redeeming qualities we may find in Deakin, Barton or whoever). And it was this new, hierarchical, racially based and teleological thinking that provided the justification for the exclusion of Chinese and Aborigines from our newly conceived nation.

Social Darwinism (which has nothing to do with Charles Darwin or contemporary "Darwinian" evolutionary theory) was a delusion that the English gentleman (or at least the educated British male) was the apogee of evolutionary achievement. Not surprisingly it held extraordinary appeal for the poorly educated, self-made Europeans of the colonies. Of course, they believed that the ocular proof of the theory was all about them. They had defeated and replaced the inferior races, whose last remnants lay dispirited and often drunk in the gutter before them. Such last survivors of the Aboriginal dominion were "the dying race", whom evolution had not fitted to compete against the world's best. And there was no guilt in their extermination. It was simply a matter of following the way of nature, and the fate of the blacks, who were just half a step above animals anyway, was not something to lose sleep over.

Other races lurked on Social Darwinism's imaginary evolutionary ladder, just a few rungs down from the whites, and these, according to nineteenth-century reckoning, *were* worth worrying about, if only to keep

them out. Foremost among them were the Chinese, whose industrious-
ness, success in business and willingness to undercut European workers'
wages were widely known and feared. Their arrival in Australia in large
numbers in the 1850s, during the gold rush, triggered a violent reaction.
As you might guess from Cunningham's observations, Chinese (mostly
from Fukien) had been in the colonies from the very first, and in the early
days of the goldfields Chinese diggers were accepted without comment.
They had been living in Australia for some while, having arrived as traders
or sailors, and were already considered part of the community. But the
news of the stupendous discoveries in Victoria sparked a tsunami of immigration from around the globe. The Chinese began to arrive in their
thousands, most of them hailing from Guangdong Province. By the mid-
1850s, most rural Chinese, who might not have heard of Rome or Paris,
knew all about Melbourne. *San Gum Shan* they called it, meaning "second
gold mountain" (California being the first), and these new arrivals seemed
almost as alien to the Fukien Chinese as they were to the Europeans.
Resentment among the Australian Europeans sparked up before too long
and soon gave rise to Australia's first wave of prejudice that was actually
ratified in an institutionalised form.

In June 1855 the Victorian government brought forward the first explicitly racist legislation to appear in Australia: it limited ships to bringing
in one Chinese person per ten tons of registered tonnage and charged a
poll tax of ten pounds for each Chinese landed. Even harsher measures
were mooted, which would have seen Chinese resident in Australian
deported. As with the Hansonite imaginings of Aboriginal cannibalism,
the supporters of this legislation created fictions to support the insupportable actions their inflamed prejudices led them to. Governor Hotham, for
instance, opined that in the vicinity of the Chinese, "our youth on the
gold-fields will be trained in vice and profligacy, and the moral growth
of the colony blighted". Hotham's nineteenth-century code-words for
buggery and gambling might well have alarmed the prurient populace,
but it's hard to see how he can have believed his own lies. There is no

evidence that Chinese who came to the goldfields were any more inter-
ested in sodomy than any other group in Australia. And for the thief
colony to have been worried about importing such vices was a case of the
pot calling the kettle black. After all, the very first European crimes pun-
ished in Australia were sexual ones involving whores, carpenters and
cabin-boys dressed in petticoats, doing exactly what is anyone's guess.

The dogma of race fostered by Social Darwinism allowed Hotham's
moralistic innuendo to go unquestioned, something less likely to have
happened in the Australia of thirty years earlier. There was also massive
support for the anti-Chinese propaganda from the growing labour movement. So hysterical had the feeling against the Chinese grown in Australia
by the latter part of the nineteenth century that in June 1888 the state premiers held a conference on Chinese immigration in which they mapped
out new, restrictive laws that all but strangled any Asian immigration to
the continent.

At the time of Federation, out of a non-indigenous population of just
under four million, more than 80 per cent were native-born, and most
were of a decidedly insular, self-satisfied mind-set. Australia was stewing
in its own juices, and it's a process that tends to continue, with some
exceptions, for the next fifty years. Remember that for the first half of
the twentieth century the overwhelming majority of Australians would
never leave their island home, or would do so only to fight in foreign
wars or to "go home" to mother Britain. We were brewing up an inward-
looking society, fearful of its isolation from Europe and its proximity to
Asia, where the fads of the West would eventually find their way to us to
become crazes only after their appeal had faded elsewhere.

Rather than encourage the nation's sense of independent worth, the
orientation of Australian society at the time made things far worse. It was
derivatively British, yet every time Australians "went home" and were
called "bloody colonials" they knew their culture was not the real thing.
Worse, Australia was steadfastly ignoring its own past and its relationship
with the land. It was the sort of place in which Manning Clark could

pen his epic history of the nation, wherein the Europeans played out their European dramas on an Australian stage that the reader hardly ever saw, like so many cartoons of European pomp and glory, caricatures on a native landscape no one could envision. That grand history might as well have concerned the fate of a colonial people in Venezuela or Zimbabwe or Alaska, so little did it acknowledge the environmental stage upon which the actors strutted. Ecologically, this was *terra nullius* indeed.

Yet this society was not static or inactive, for one of the defining characteristics of Federation Australia was its can-do attitude. If the stumps of vast, felled forests littered the wheat fields, a plough could be designed that jumped over them. If most of Australia was a desert, then the rivers could be turned inland to water it. Some, like Edwin Brady, even argued that Australian deserts were only deserts in appearance. A providential deity had stored vast volumes of water underground in the Great Artesian Basin for the use of the white man. Australian deserts would come to bloom with crops to feed hundreds of millions, and everything was bent on the hope of a great nation. Indeed, everything in Australia was called Great in advance, in the desperate expectation that it should be so: the Great Dividing Range, the Great Australian Bight, the Great Artesian Basin, the Great Sandy Desert. How could it not be a great country – an Australia Unlimited? How not indeed? And what did these proud hopes feed on? What soil did they come out of? Self-congratulation, the prejudices of insularity, self-delusion.

Just a hundred years after British and Aboriginal cultures came face to face on that Sydney beach, reaffirming the essential fact that humanity was one, the new Australian society had lost sight of the one great truth it had been allowed to glimpse. No one wanted to look backward to the Dreamtime, or to the historical dream where blackfella meets First Fleeter, they wanted to look forward to an Australia triumphant – an Australia built not on communion but on the dispossession of others. For those colonial Australians the stupendous success of their first century was nothing compared to what they believed was yet to come.

They affirmed the faith that their nation would grow to be hundreds of millions strong – all white and all British, of course – a mighty outgrowth of the Anglo-Celtic peoples, destined to become the superpower of the south. A United States of Australia with its own manifest destiny.

Australian schoolchildren are taught that the colonial period ended in 1901 with Federation. Two years ago we celebrated that supposedly momentous event; yet for all the millions spent by the federal government on exhibitions, television programs and other celebrations, very little enthusiasm or sense of public ownership of the centenary was evoked. People, especially politicians, seemed discomforted and dismayed, even vaguely embarrassed by it all. Why? Well, there are several important reasons for such a lukewarm response. One is the many Australians who harbour a lingering resentment over what they see as the hamstringing of the republican movement by the Prime Minister; they are damned if they'll celebrate *anything* with the Howard government. Their refusal to join the celebrations was no doubt reinforced by their feeling that our constitutional evolution is not yet complete, and that a true Australia will only come into being when we shed the British monarch as our head of state.

All of which is plausible if not self-evident, yet there's something deeper at work. Perhaps the centenary was a fizzer because, in the larger play of Australian history, Federation did little for Australians. I say this because I do not believe that Federation ended Australia's colonial status. It was an unmitigated disaster for non-European Australians. Not only had the state premiers choked off Chinese immigration, but anti-Asian immigration policies would become a keystone of the new federated nation. And Aboriginal Australia, the poor fellows of our country, also had a lot to mourn in 1901. For decades the Aboriginal people of South Australia, that forward-looking state, had exercised their right to vote, but this was taken from them at Federation through the coercion of the other states, in a deliberate and cynical curtailment of fundamental human rights. Indeed, in 1901 Aboriginal Australia was firmly and deliberately locked out of the new nation. As I look back on those old paintings of proclamations made by men in top hats, Federation looks more and more like a game in which colonial Europeans re-arrange the deck-chairs on a Titanic full of self-satisfied fools, unable to see the icebergs that their whole ocean liner of complacency is inexorably sliding towards. Some of those icebergs are social and some are environmental, but they are all enormous. Compared to them, how much of a chance does that ship have?

THE COLONIAL DRAIN

I tend to think of the forty years either side of Federation as the crisis time in the deadly plague we call colonial Australia. Either the patient would succumb, or else show some faint signs of recovery. And it's true that great tracts of Australia's ecology came perilously close to death. Around the time of Federation, foxes and rabbits were spreading like a plague across the landscape. Both had been released in Victoria in the mid-nineteenth century, their introduction an integral part of the dream of creating a second Britain in the antipodes. Their impact was to be massive and massively damaging. In 1901 some European was afforded a last glimpse of the curious pig-footed bandicoot, and in the forty years following its extinction one species of native mammal after another would disappear, until a boggling 10 per cent of all native mammals was lost, giving Australia the world's worst record in modern times for the destruction of natural heritage.

Despite the terrible nature of this loss, hardly a concerned voice was raised in the Australia of the early twentieth century. Instead, until the 1930s there was a bounty on endangered marsupials in Tasmania, Queensland and elsewhere, while unregulated markets for marsupial skins in South Australia were all but eliminating the few remaining natives. We were helping smooth the pillow of our dying marsupial fauna and making a little money out of them at the same time.

And it was not just mammals that were threatened, but the very soils and waters of the continent. And, yes, it was all done for the best of reasons – that old weakness for, that old faith in, beautiful lies.

Of all the lies spawned by our nation, none has been as damaging to the Australian environment as the overweening lie of boosterism that prevailed around the time of Federation, and which indeed still prevails in some quarters today. A vast series of schemes was propounded, one of the most grandiose of which was George and William Chaffey's proposal in 1887 to irrigate the Murray-Darling Basin. The Chaffeys were Californians, and they planned to bring American-style irrigation Down Under. The governments of the day were delighted and granted them the right to pump as much water as they wished, *gratis*, along with a gift of 50,000 acres and the right to purchase another 200,000 at 17 shillings per acre.

If the irrigation scheme was to succeed there had to be water in the river to feed it. This was a problem, because the Murray, like most Australian rivers, ran on a boom and bust basis. In some years drought reduced it to a mere trickle, but at other times La Niña would turn it into a majestic waterway that could burst its banks, so that paddle-steamers could moor at flood-beset shearing sheds to load their bales of wool. Because the river had run like this for millions of years, the species that lived in and around it were adapted to these particular conditions to such an extent that they needed them to thrive. With artificial irrigation and the consequent increase in river trade, a series of locks was installed along the Murray based on a 1915 agreement, impounding its water, damping down the effects of drought and flood and, in ways that should have been predictable, dramatically diminishing the river's once rich abundance of fish and bird life, its whole biodiversity. Where waterbirds once bred in their tens of thousand and Murray cod grew to the size of groupers, today there is nothing but carp and salt scalds.

As the Murray was being drowned in its own water, politicians everywhere were looking for cheap ways to clear the dreary Australian scrub. Sir James Mitchell (whose enthusiasm for establishing a Westralian yeomanry saw him dubbed "Moo-cow" Mitchell by the populace) had a plan just after the Great War to import 75,000 British migrants to Western Australia who would clear the karri and jarrah forests in order to establish a dense network of tiny farms. The place would then look like Europe, its primeval forests vanquished under a blanket of bucolic English bliss. But it didn't work. In the face of the truculent Australian soils, and the sheer weight of forest to be disposed of, the scheme collapsed.

Government was always at hand to fund ludicrous and damaging projects, and there was no shortage of money for them. Among the most tragic of these futilities were the soldier settlement schemes, which saw brave men returning from the horrors of war to become the shock troops in Australia's war on its own environment. They were given grossly inadequate plots of "scrub" to set up their futures, and after they broke their hearts clearing it and grubbing up mallee roots, they drifted away, as surely as the exposed sand of their newly cleared fields took its own journey eastwards on the relentless drought-winds.

Through all of this the boosters never seemed to learn. Melbourne was only sixty-six years old at Federation, yet it had already seen several cycles of mad speculation followed by deep recession, each of which had been accompanied by relentless environmental vandalism. Fortunes were certainly made by those grand men in their gold-trimmed robes whose portraits still line the walls of the Town Hall's Yarra Yarra Room, but the cost was terrible, and nowhere was the environmental toll more evident than in the Yarra River itself.

When John Batman sailed into Hobson's Bay, he found a paradise, a temperate Kakadu whose waters thronged with waterfowl. Black swans dotted the bay in their countless thousands, while magpie geese and brolga bred along the banks of a limpid Yarra. Within a decade the Europeans had shot most of the waterbirds out of existence and turned the river into a slaughterhouse-lined sewer. The tallow works were surrounded by piles of bones more than ten metres high, and the edges of the waterway were strewn with guts, amidst which pigs wallowed. For a century the jewel in Melbourne's natural crown would remain a putrid drain, shunned by wildlife and Melburnians alike. The fate of the Yarra was a foretaste of how Australia's colonial people would treat the water throughout this driest of continents. It was for us essentially a thing to squander and lay waste. The nation's recklessness with its waterways was to reach its apogee in the decades following the 1950s, and the greatest catastrophe of them all was the Snowy Mountains Scheme.

Fifty years ago Australian hearts swelled with pride at the mention of the Snowy scheme. It was the largest engineering project the nation had ever undertaken — our first attempt, in earnest, to realise the dream of turning the rivers inland and watering the deserts. The project was inspired by the works of the US Army Corps of Engineers — "Build me a dam just like old Uncle Sam," sang the buxom young at the time. The Corps' work of straightening, damming and concreting rivers in the US was so extensive that today there is hardly a natural catchment left in the lower forty-eight states. Marc Raisner's *Cadillac Desert* documents the environmental and financial costs of this ill-conceived work, which was copied around the world with tragic results.

The Snowy scheme can store 7,000 gigalitres (7,000,000,000,000 litres) of water, which can be released to the Murray River when needed. A flow able to be released at will was a potent weapon in the scheme that would all but kill the Murray, but the environmental cost for the Snowy River was even worse. It resulted in the reduction of the once mighty headwaters of the Snowy to a trickle, because 99 per cent of the river's flow had been diverted. The water allowed for massive investment in irrigation projects downstream on the Murray, and in its turn that extra water poured on the land played a significant part in the salinity crisis.

Only one politician saw any danger in this much-touted scheme. Sir Thomas Playford, the cherry-growing Premier of South Australia, argued with Sir Robert Menzies, "Now that they know there will be additional amounts of water, I have not the slightest doubt … that they will develop their use of water much more extensively," he said. "I do not accept for one moment the argument that because more water will be coming in, there will be less likelihood of restrictions." Playford foresaw that the irrigators of the Murray-Darling Basin would become addicted to the "free" water of the scheme. Today that addiction is as debilitating as it ever was, and it has made it exceedingly difficult to improve the Murray's ecological lot.

Although the consequences of the Snowy scheme were widely acknowledged to have been disastrous, the fiftieth anniversary of its inauguration was celebrated exuberantly. But to continue to pretend that the Snowy scheme was a great success, a noble and ennobling enterprise, is to perpetuate a lie. Quietly, at the same time as the celebrations, there began attempts to reverse the scheme's effects. Agreement has now been reached for 28 per cent of the scheme's water to be returned to the Snowy River. Most of that water will come from increased efficiency of water use on the Murray. This might sound like good news, but, as we will soon see, nothing is so simple: increased water efficiency poses another profound threat to the Murray.

By the 1960s, some Australians were becoming concerned at the perilous state of their environment. As biologists searched in vain for creatures who had been abundant only a few decades earlier, the full extent of the extinction crisis began to be realised. Inspired by Rachel Carson's classic *Silent Spring*, people began to see the damage done by chemicals such as DDT, and when they looked at their urban waterways, they found dead, open sewers. But the early stages of environmental awakening did not immediately translate into the most logical plan to rescue the Australian environment. No, there were more beautiful lies to contend with.

Perhaps the most influential book in the early days of the Australian environmental movement was Jock Marshall's *The Great Extermination: A Guide to Anglo-Australian Cupidity, Wickedness and Waste*. Marshall was a one-armed rebel – always at odds with the establishment, even after he became founding Professor of Zoology at Monash University. He also had a wonderful way with words, and *The Great Extermination*, written in 1966, moved people powerfully, for it exposed for the first time the shocking extent of biodiversity we had lost in Australia. It also flew in the face of economic and political orthodoxy; at that time forests were for timber, whales were for catching, and some of our most powerful politicians won wealth and influence clearing "the scrub" with ball and chain. Even worse, some of the most beautiful places in Australia – Lake Pedder, the Gordon River and Frazer Island – were under threat of destruction.

Inevitably, political action crystallised around a few flagship issues. One was the battle to save those special places, and through the 1970s such campaigns met with increasing success. Other environmental success stories included the preservation of "wilderness" areas, the protection of native species such as the koala, the cessation of whaling by Australia, and the saving of large tracts of forest. From the perspective of a contemporary environmentalist, however, this list of victories looks strange, because it does not focus on the key issues of land, soil and biodiversity conservation. In fact, it makes sense only if one understands the philosophy of the early conservationists. Encouraged by books like Marshall's, many believed that everything would be fine if only the dead, white hand of the colonial – male – Australian could be lifted from the environment.

It is worth examining each of the goals of the early conservation movement separately, as each has had a significant effect on the world we live in today. Foremost among them was the battle to save Australia's threatened "sacred" places. Such campaigns have been splendidly successful in preserving various geographic locations from deleterious development, the stand-out victories here being Frazer Island and the Franklin and Gordon Rivers. Over time, however, the program widened into a struggle to preserve at least one representative sample of each of Australia's environmental types. Typically, the goal is to "save", in the form of a national park or reserve, 10 per cent of each type of environment. This is fine as far as it goes, but as a strategy to conserve biodiversity it has been markedly less successful, and the reasons for this failure of practical conservation go back to the original sin of terra nullius.

If we look around our national parks today, what we see in the great majority of cases are marsupial ghost-towns, which preserve only a tiny fraction of the fauna that was there in abundance two centuries ago. A classic example is Royal National Park south of Sydney. It's the nation's oldest national park, yet over the last few decades it has lost its kangaroos, its koalas, its platypus and greater gliders. Clearly, it is a fallacy to believe that proclaiming more such reserves will do very much to preserve Australian wildlife.

Even though I knew this in my heart, none of it dampened my euphoria when, as a young environmentalist, I heard that another bit of Australia had been "saved" in a national park. Weirdly like a young British imperialist at the height of empire admiring the expanse of pink on the globe, I'd examine maps of Australia and inwardly cheer at the proliferating green areas. It was only when I gave away the falsehood of terra nullius (and the accompanying belief that the land was "ours" to carve up according to our inclinations) that I began to think differently. As you may imagine, the wilderness movement has faced similar difficulties, and even in strict wildernesses the plight of wildlife is just as dire. The hard truth is that the philosophy on which these movements were based was flawed, and there was no magical cure when European interference with the land was done away with. In fact, in many instances, things got worse.

This dispiriting failure was one of the strangest manifestations of *terra nullius*. The environmentalists for all their good intentions singularly failed to appreciate that the land had been managed – indeed carefully shaped – by 47,000 years of Aboriginal occupation. Unless you've had the good fortune to have spent some time with hunter-gatherers, it's difficult to comprehend the depth of the impact they have on the environment. It's also difficult for anyone to comprehend the meaning of 47,000 years. To take the first point, it is important to realise the unique impact human inhabitation makes. A collection of large, intelligent omnivores can consume an astonishingly wide variety of food. Aboriginal Australians gave paramount importance to two things – hunting and fire. As hunters, the Aborigines were the primary carnivore on the continent, taking the lion's share of animal protein. Through the use of fire they were also the top herbivore, with their fire consuming more vegetation than any herbivorous species, perhaps more than all the vertebrate herbivores put together.

The black hunters culled the wildlife of Australia and preserved what was viable. So extensive and effective was Aboriginal hunting that the large marsupials were rare at the time of European contact. It was a red-letter day when the explorer Sir Thomas Mitchell saw a kangaroo, and in the 1840s the naturalist John Gould felt that both red kangaroos and koalas were so rare that they were doomed to extinction. It is also astonishing to read explorers' journals and discover how often they killed a kangaroo or a seal only to find a broken-off spear tip in the creature's body.

Hunting and fire shaped the Australia we imagined we had discovered. We have already seen how Aboriginal fire might have influenced one species, the white-footed rabbit-rat, but the truth is that it had a profound impact on all burnable vegetation types in Australia. Indeed, if we look back on the fossil record, it's not an exaggeration to say that Aboriginal fire and hunting literally made the Australian environment that the Europeans first encountered. It was a vast, 47,000-year-old human arte-fact, designed to provide maximal food and comfort to its inhabitants in the most sustainable manner.

When Aboriginal fire and hunting ceased, the environment began to change. Over the past 200 years south-eastern Australia has seen bushfires get ever more devastating and frequent – which is just one reflection of that change. Others include the population explosion among the larger kangaroos and koalas. None of these could be described as in any way endangered today. Along with these changes, a series of pestilences was let loose by the Europeans. Among the animals, the foxes and rabbits were the worst, and their effect was exacerbated wherever the farmers had exterminated the dingo. But hundreds of other species, from plants to carp to cane toads, have all altered the places where they proliferate. To think that we can walk away from managing the new environment that this history has created is a form of madness.

It's in this context that we need to examine another focus of the Australian conservation movement – the preservation of native species such as kangaroos and koalas. It's to this cluster of causes that those inter-ested in the prevention of animal suffering or cruelty have gravitated, some of them with exceptional dedication. Several years ago I spent some time with a group of people who had come together to care for koalas that had been badly burned in the Port Macquarie bushfires. Among them was a knockabout young man who risked his life to climb the trees the injured creatures had taken shelter in. But by far the largest number were women whose families had grown up. One remarkable woman, a retired nurse in her late seventies, stayed awake for twenty hours at a stretch tending a koala that had been terribly hurt. Its claws, ears and most of its nose had been burned away, yet it lay there stoically as its carer changed its bandages and anointed it with creams. A more highly strung animal would almost certainly have died of shock from its wounds, but the small brains and lethargic metabolisms of koalas serve them well in such stressful circumstances.

Wildlife workers as devoted as these certainly do no harm to the species they try to protect, and the individual koalas they treat benefit greatly. Whether koalas as a species are endangered or not is irrelevant to them. They see a suffering animal and try to help. The picture is not quite so clear, however, when it comes to those well-wishers who try to "save" the kangaroos. The four large kangaroo species of southern and central Australia (the red, euro, eastern and western grey) support a substantial and sustainable meat industry, and it is this industry that is the focus of the anti-culling activists' outrage.

The battle to close down the kangaroo industry has been protracted and bitter. Whenever a new opportunity is seized by the industry, whether nationally or internationally, its opponents spring up and do their best to close it off. This is really a great pity because from an environmental perspective the kangaroo meat industry is by far the best managed meat-producing industry in Australia. It is regulated by the federal government, which sets a quota annually that is strictly controlled. The on-the-ground evidence of its success is also abundant, for no one could argue that the target species are in any way threatened today. Environmentalists should actively support the industry. In fact, many of them do. The most bitter opponents of the kangaroo industry are drawn from the animal welfare and animal liberation lobbies.

Even from an animal welfare perspective, however, opposition to the controlled killing of kangaroos seems ill-founded. The first a kangaroo knows of its interaction with the industry is a spotlight and a bullet to the brain, which usually results in a clean kill (animals shot elsewhere in the body are not marketable for meat). What is suffered by domestic stock — castration, de-horning, road transportation and death at an abattoir — seems barbaric in comparison. Yet it's the kangaroo industry that finds itself the target of the animal welfare lobby rather than farm and abattoir managers, a fact that perhaps can only be explained by the "cuteness" of the "wild and defenceless" marsupials. The pathos of the situation is that not only are the campaigners not serving their own best interests, but they are threatening, however unwittingly, to damage the environment at the same time. To shut down the kangaroo industry would be to cut off one of the most sustainable and humane animal-based industries in rural Australia.

The whale is another charismatic mammal that has received great attention ever since the early days of the environment movement, and Australia's anti-whaling campaigners have been exceptionally successful. Australian whaling ceased in 1978 (with a legal ban following in 1980). Pressured by the green lobby, Australia has become one of the staunchest supporters at the International Whaling Commission of closing the entire global whaling industry – a fact that will not bear scrutiny. It's true that in contrast to kangaroos and koalas most of the larger whale species were on the brink of extinction when campaigns to halt their killing commenced, but today the smaller whale species are abundant enough to allow for sustainable harvesting. In order to maintain the blanket no-kill approach, the campaign to "save the whales" has had to depart from a strictly environmental logic.

In practice anti-whaling campaigners claim that the entire whaling industry should be shut down because, as a form of multi-nationalism on the high seas, it cannot be regulated. The same logic would close down every blue-water fishery, yet little green energy is expended on this. In fact, because the whaling industry is so comparatively well managed, good environmental managers should be calling for the closure of a lot of other fisheries before they even begin to think about whales. Nor does it help that they spread the message that whaling is morally wrong because whales are sentient beings like ourselves – an extraordinarily successful gambit because the public has been led to believe that whales are not far behind humans on the intelligence scale. It is just another lie.

What people fail to realise is that the *Cetacea* (the group to which whales and dolphins belong) is an extraordinarily diverse order of mammals. It includes relatively large-brained hunters like dolphins and killer whales (which have the demonstrable intelligence of land-based hunters such as dogs) and tiny-brained filter feeders such as the blue whale. These leviathans are aquatic vacuum-cleaners, whose need for intellectual power is slight indeed. Dolphins and killer whales have never been part of the commercial harvest. It's the small-brained (relative to body-size) filter feeders that are hunted by the Norwegians and Japanese. If these animals are closer in intelligence to the sheep than the dog, is it morally wrong to eat them if they can be harvested sustainably? My view is that at present the anti-whaling lobby is frustrating the attempt to develop a sustainable industry based on these creatures, and is therefore frustrating good management of marine resources. It is a sorry evolution to turn from being at the forefront of marine conservation to being opponents of best conservation practice.

The details of the history of whale conservation and its consequences tell an interesting and complex story. There is the fascinating return of the southern right whale, whose proliferating numbers today support a vigorous whale-watching industry. By the 1840s, southern right whales had been hunted to near extinction in Australian waters, and they remained almost entirely absent from our shores for the best part of 130 years. Then in the 1970s they made a comeback. They're easy to see when they're about because females come right into the beach break to give birth and shelter their young from sharks. They visit the coast as far north as Sydney, and last year one mother braved the traffic of Sydney Harbour to bring her calf right up to the Opera House and Harbour Bridge, where the pair cavorted for some days.

Many believe that the recovery of the southern right whale proves the efficacy of the anti-whaling campaigns, but nature is rarely that simple. The truth is that no one knows why, so many years after it was last hunted on any scale, the species began to recover. It is possible that the occasional one may have been killed by whalers looking for other species, and that in a tiny population this may have prevented a resurgence; but scientists have also pointed to another, more intriguing possibility. By the 1960s and '70s the larger whales were becoming rare, and whalers turned their attention to the abundant but smaller minke and sei whales. These species seem to have flourished as their larger cousins went into decline, perhaps because of an increase in the amount of krill available to them. It was only when whaling began to deplete these smaller species that the numbers of southern right whale began to increase. It seems possible that once the minkes became abundant, the few remaining southern rights could not compete with them for food, and it was only as the minkes' numbers were thinned by hunting that the southern rights found "breathing space" to increase their population. If so, then it appears that the whalers saved the southern right whale, not the environmental campaigners. Whatever the truth of the story, it demonstrates the complexity inherent in managing natural systems. Even the kindest of intentions can lead to disasters, and in my opinion only the best science, and a commitment to the policy that makes best sense in the light of it, is sufficient safeguard against unintended blunders.

The issue that most dominated Australian environmental politics in the closing decades of the twentieth century was the campaign to "save the forests". Throughout the 1990s it consumed the time of Ministers for the Environment and their advisers like no other, and it has been claimed that forests were the decisive issue in the 2001 Western Australian state elections. Although many aspects of forestry practice offend anti-logging campaigners, at the heart of the matter is the fate of Australia's last unprotected "old-growth" forests. Their trees grow in regions like Tasmania's Valley of the Styx, where the mountain ash soar more than eighty metres overhead. Such forests are magical places for many people, and it seems sacrilege to fell them for construction timber or, even worse, wood chips. Yet in terms of biodiversity and soil and water conservation, the fate of these last unprotected stands of old-growth forests is relatively unimportant.

While I admire the achievements of the anti-logging campaigners, I must admit to being irritated by the vast amounts of time and money that have gone into the campaigns. No one could claim that Australia is not much the better for setting aside these forests. Yet compared with the other issues that could have been furthered if they had received as much attention, it all looks like a lost opportunity. It is an unhappy fact that if salination or biodiversity loss, for example, had attracted half the funds and attention lavished on the forests, we would have seen far more dramatic improvements in the overall environmental health of Australia.

So how are we to characterise the environmental movement of the late twentieth-century? It's vital to see it in context. It arose during an era when government and private enterprise were thoughtlessly destroying some of the natural world's most precious assets, and it formulated strategies as a reaction to that threat. The movement began at a time when scientific knowledge about the extent of the environmental crisis and its true nature was rudimentary. There was not the widespread awareness at the outset that we are engaged in a struggle for our very futures with overwhelming and fundamental issues of soil, water and air quality. Many campaigns did a fine job in achieving their principal goals or in raising public awareness, but then found that as the targets of their wrath (whether it was forestry, whaling or the kangaroo industry) began to adapt in key areas, or even lead the way in environmentally sustainable practice, they had to shift ground. And all too often, that shift has been away from a focus on sustainability and towards subjectively emotional issues such as animal rights and tall trees. These are not bad things in themselves, but the way they have been used has often, with the best will in the world, actually run counter to good environmental management.

What then of the real and fundamental issues – land, water, air and bio-diversity conservation? It is in this realm that our future will be decided. And I have no hesitation in saying that we must approach these issues on a war footing, for we are in a struggle to the death with the threats which are posed to our well-being as a nation. Many of the perils we must deal with crystallise at this moment in time around our rivers, for that is where most of the nation's agricultural wealth is won, where just 1 per cent of the land area produces 98 per cent of its yield in financial terms.

Water

Ever since the first of the Chaffey Brothers' pumps began sucking water out of the Murray River over a century ago, the obsession has been water. It's liquid gold to a primary producer, and access to it is the core require-ment of any successful rural business. As a result, today 80 per cent of the median water flow of the Murray-Darling Basin is extracted for irrigation and other consumption.

Our history of trying to cope with Australia's mounting water crisis reads like a tragedy with a natural curve towards catastrophe. We have known for decades that something must be done to prevent this, yet each move we have made seems to have made the situation worse.

One of the most damaging initiatives was the establishment of a water market. The concept derived from the economic rationalism of the 1980s, but it has had the unintended consequence of escalating the price of licences. So bad has this unintended fallout become that a water licence that sold for $30 in Moree, New South Wales in 1970 today trades for $1.4 million! Little wonder that Australia's mega-rich have rushed into the water licence game. This has made it formidably expensive for any government to pursue water reform and, not surprisingly, it has estab-lished in the minds of the purchaser that they actually own the water they have paid to extract.

In 1997, the extraction of water from the system was pegged at 1993–4 levels. The problem is that this level of extraction is simply unsustainable. Indeed, in many parts of the Basin we have the Alice in Wonderland situation where more water has been allocated to irrigators than actually exists. Discussions are now being held to consider returning 20 per cent of the water used for irrigation to natural flows. This will be a vastly expensive exercise, and the problem is so massive that the chances of it by itself substantially restoring the catchment to ecological health are slim.

The difficulty boils down to defining what rights an irrigator holding a water licence actually possesses. The system is so chaotic that, at the present time, increases in water-use efficiency by irrigators actually result in less flow in the river system. That's because highly efficient irrigation results in more water being used by the plants (instead of the water running back through the ground table to the river) and because irrigators are allowed to keep the allocations liberated by their increasing efficiency. Clearly, such increased efficiency stops pollutants like salt from reaching the river, but this comes at the cost of less water being available in the river, which is an intolerable situation.

If we are to succeed in our battle to save the Murray-Darling Basin for the use of future generations of Australians, we need a total re-think of water rights. The Wentworth Group has suggested that the framework for water rights reform should be drawn from the concepts appertaining to the Torrens title for land, combined with the idea of shared ownership enshrined in company legislation and informed by innovations in internet banking and risk assignment. The basic right should be to use a proportion of available flow for a finite time. Essentially the group proposes that each water user should receive a proportional share of the resource with a clear statement about its reliability (or otherwise) and the risk of change without compensation. The water should also be returned to the river in the same condition as it left it, with pollution licences required of producers who release sediment, salt and unwanted nutrients back into the river. Good farmers, adhering to best practice, would pay nothing under such a provision.

The Salinity Crisis

Intimately tied up with the issue of water rights is the blight of salination. It has different causes in different environments, but at its heart is the fact that European agricultural crops are inefficient. They don't use all the water that falls from the sky in the way much of the native vegetation does, and some of the water soaks down into the soil profile, dissolving vast amounts of salt that have lain as crystals in the soil since time immemorial. The salt then joins the water-table, which rises, killing everything in its path. Not surprisingly, if water is poured onto the soil by irrigation, the problem is made many times worse. In the Murray the situation is exacerbated by the high salt content of the rocks and sediments underlying the floodplain. The river's very restricted outlet to the sea also magnifies the problem, because the salt (and all other environmental pollutants poured into the river) keeps getting re-deposited in the rich floodplain soils rather than being flushed out to sea. It has been calculated that the salt and other pollutants we release into the river today will take centuries to flush out.

The problems of the Murray are clarified by looking at the situation in Western Australia, where the most precise studies have been undertaken. There, the difficulty was caused by clearing one of the most bio-diverse vegetation types in the world – the kwongan and related scrubs. These wondrous vegetation communities harbour over 4,000 species of native plants and are considered one of the globe's hot-spots of biodiversity. Anyone who has travelled to Perth in the springtime would have seen the fantastic display of wildflowers it produces. The kwongan (an Aboriginal word) was so efficient at using water that hardly a drop escaped its roots – everything was used by the plants. And this is a good thing, for under every square metre of land in the wheatbelt lies between 70 and 120 kilograms of salt crystals. If dissolved and brought to the surface, it could turn much of the region into a convincing simulacrum of the Dead Sea.

Such a development would have serious economic as well as environmental consequences, for although the Western Australian wheatbelt produces only a fraction of Australia's wheat, it plays a vital role in ensuring market reliability. That is because eastern Australia is subject to dramatic fluctuations in yield brought about by drought and flood, whereas the wheat produced in the much more reliable west allows us to fulfil contracts in drought years and to be seen as a reliable trading partner – valuable assets in the world of global trade.

From the very earliest days of settlement, people in Western Australia noticed that clearing the native vegetation made creeks salty. As early as 1905, W.E. Wood, a railway engineer, informed government that railway water supplies were being made too salty for boiler use as a result of land clearance. In 1924, as the push to clear the scrub gained momentum, Wood published his results in a landmark scientific paper, the political response to which was appallingly ignorant and dismissive. Moo-cow Mitchell, then leader of the State National (Rural) Party, captured the feeling when he quipped, "I am afraid that if the good Lord had provided scientists when Adam and Eve were created, no useful work would have been done."

So determined was the state government to go ahead with development that for decades parties of all political persuasions ignored the advice of their own experts, experienced farmers among them, and pushed ahead with extensive land clearance. After 1945 the largest land clearance program in Australia for more than a century commenced – the soldier settlement scheme. It was taken a step further in the west in the 1960s by the "Million Acres a Year Scheme", instituting an even more wholesale clearance of native vegetation. Today, the result of this greed and stupidity is simply terrifying, as a few figures will suffice to demonstrate.

Two and a half million hectares of agricultural land are salt affected, 1.8 million of which are in the southwest of Western Australia. A further

10 million hectares are at risk nationally, 6 million of which are in the southwest of W.A.

The Western Australian wheatbelt suffers from one of the worst examples of dryland salinity in the world: 30 per cent will be affected in the next few decades. Over forty wheatbelt towns face damage to buildings, roads, railways and airstrips. The cost to transport infrastructure alone in these towns is estimated at more than half a billion dollars. Salination in the southwest is already resulting in massive human hardship, including loss of farms, suicide and the destruction of Aboriginal heritage including sacred sites and significant vegetation, old campsites and archaeological deposits.

The southwest region's few areas of natural vegetation (90 per cent has been cleared) are threatened by salination, with no solution in view. Over 850 unique plant species are threatened.

By 2100, more than 43,000 kilometres of the nation's rivers are projected to have elevated salt levels, including most if not all rivers in the southwest. In the wheatbelt today, rivers are fifty times more salty on average than they were at the time of settlement, and it will take more than 1,000 years to leach the newly released salt from the land.

The problem of salination in Western Australia is in fact so immense that many experts, including some conservationists, are suggesting that it is insoluble and that we should simply walk away from the mess we have created. Many others, however, are fighting the salt for their livelihoods and cannot or will not walk away. Around half of all Western Australian farmers had become involved in Landcare by 2000, and re-vegetation and some engineering projects have commenced. Their likelihood of success, however, is highly uncertain, with some concerted plantings – undertaken specifically to protect endangered native vegetation – having already failed. "No broad-scale solution is presently [sic] available that is simple and cost effective," says one expert. A salinity tax may be the only way of raising the astronomical funds needed to begin to tackle the problem, but many Australians see salination as so remote from their experience that such a tax is unlikely to be accepted.

The Murray-Darling Basin

Now let's return to the Murray-Darling Basin, where 40 per cent of the nation's agricultural produce is grown. Although the figures are not known as precisely as in the southwest, the scale of the problem is equally terrifying. To quote just one alarming statistic – within seventeen years Adelaide's water supply (drawn in large part from the Murray) will be too salty to drink on two days out of five. The Murray-Darling Basin Commission has been working for decades on solutions, yet these have been slow to develop. Meanwhile, ironically, the situation has been rapidly deteriorating. A good indication of this is land clearing, which leads to salinity. It has been on the increase Australia-wide, not least in the Basin, since 1997. Today around fifty football fields of native vegetation are being cleared every hour – a rate that is only exceeded globally by Brazil, Indonesia, the Congo and Bolivia.

The goal of those who care about the Murray-Darling Basin must be to restore the river system to environmental health. As a first step, plans are being made to put water back into the river – to restore so-called "environmental flows". This means restoring the river, as much as possible, to its pre-European mode of operating – in effect undoing all the damage of the past 200 years. It's a painful initiative to take, for the water necessary for environmental flows has to be clawed back from that allocated to irrigators. Those who are able to will probably build farm dams to compensate for the loss, catching the water before it reaches the river and necessitating more unwelcome legislation if the scheme is to have a chance of success.

The scheme also requires the restoration of the forests of the upper catchment; yet in the short-term, because they are growing vigorously, they will also take water and so reduce the flow. And as environmental flows are restored, because of the salt already released, the river will run very salty in dry periods (which are necessary for river health), further endangering irrigators. Huge efforts are being put into computer modelling to predict how the highly complex ecology and hydrology of the river system will react to the changes, though so far no conclusive results have been produced. The benefits of the scheme are far from assured, but to do nothing will certainly result in the biological death of Australia's greatest river system – an event quite probable during the lifetime of some readers of this essay.

Pollution of the Atmosphere

Before we leave this necessarily gloomy passage through Australia's environmental woes, one final factor needs to be considered. Climate, along with soil and water, is one of the "great controllers" of all life, and there are clear indications that all is not well with the globe's climate. The Earth's atmosphere is about as thick, proportionately, as the skin on an onion, so it is little wonder that it is easily polluted. Concerns about global climate change have seen more than a hundred nations club together to produce, twice each decade, a report on the state of the world's atmosphere and climate. The reports are the fruits of millions of work hours on the part of humanity's finest climatologists. They are also consensus political documents, which can make for tedious reading.

The first such report, produced in 1991, was a mild affair. I perused it and decided that there was little to worry about, for the present at least. The second report, produced in 1996, was slightly more arresting. It predicted that climatic change in the order of that seen over the past 5,000 years might occur in the next hundred. If this were to happen, cyclones might ravage Australia as far south as the New South Wales coast, and tropical corals would sprout in Sydney Harbour.

The third report, released in early September 2001, was very different again. Reading the executive summary I was filled with alarm, and delving further brought greater apprehension. The report for the current half-decade predicted a far more massive change than any hitherto – a warming of between 3 and 7 degrees Celsius over the next 100 years, with a 6 degree rise being most likely. Here was change on a scale that had not occurred since our species first evolved on this planet. And my grandchildren, if I'm fortunate enough to have any, will see it. The report should have made headlines world-wide. Instead it was buried under the cascade of news from the events of September 11.

It is almost certain that global climate change is already being felt in myriad ways. Rainfall in the southwest of Australia has been down 20 per cent on average for each of the last twenty years; climatic events are becoming more extreme; and various species – including diseases – are already changing their distributions in response to the warming. This is just the tiniest taste of what is to come. And it is we who are causing it. Every time we drive the car, turn on the power or buy manufactured goods.

Like salinity, solutions to the problem of global climate change are not easy to come by. The Kyoto Protocol is a vital beginning to the fight, but it is so unambitious that even if it were implemented tomorrow, it would do almost nothing to circumvent the coming catastrophe. The only thing that can deflect the looming disaster, I believe, is a massive personal commitment to solving the problem by the people of the West. If we all purchased a solar panel or solar hot water system, all insisted on energy efficiency in all aspects of our life, and made these technologies available to the developing world, then we might stand a chance.

The Government Response

It is worth examining the Australian government's response to the overwhelming challenges facing us in these areas. The refusal to ratify the Kyoto Protocol will almost certainly, in time, be remembered as the greatest failure of the Howard government – *Tampa*, detention camps and Iraq notwithstanding. Yet there have been a few small moves forward. Subsidies for the purchase of solar technology are still available but may vanish in mid-2003 as anti-terrorism measures eat away at the budget. Carbon trading has been introduced, and occasional rumblings are heard about the problem of water.

In terms of firm commitment – as opposed to talk – one outstanding initiative has been achieved. The mid-to-late 1990s saw the Howard government spend an unprecedented sum of money on the environment. Under the guidance of Senator Robert Hill, then Minister for the Environment, 2.6 billion dollars, in large part flowing from the partial sale of Telstra, was invested in environmental issues. At the time the spending was being rolled out, many professional environmental managers and academics voiced their strong opposition. They were concerned that community organisations such as Landcare groups and other non-scientific bodies were receiving the lion's share of the funding, and that not much in the way of environmental remediation would be achieved by them. As the program began to draw to a close, I started to fear that they might have been right, for many of the Landcare projects were small-scale and unscientific in nature, and the sum of environmental achievement did look rather meagre.

Whenever I travelled in regional Australia, however, I saw what were (for me at least) many unexpected consequences of the program. Rural people had got the message, loud and clear, that the environment was on the agenda. And even more than this, through projects funded by the program they had acquired practical experience in dealing with the specific environmental challenges facing them. It was not that their projects had necessarily solved their host of problems (whether in whole or in part), but they had produced a massive change of heart, a revolution in the hearts and minds of rural Australians. That is where the first battle must be fought and won, and the results of Hill's program in the bush so far have been first class. Partly as a consequence of the program, country Australia – once the last redoubt of red-neck environmental thinking – has vaulted ahead. Today some of the very best innovative practice in environmentalism is to be found in the bush. Now it's the environmental practices of those inhabiting the nation's vast urban feedlots – the teeming cities – that lag so dismally behind.

One Sunday morning in the mid-1980s, when my children were still young enough to wake me with the dawn chorus, I turned on the television to see our prime minister of the day, Bob Hawke, casually suggesting that Australia's population should reach 20 million by the year 2000. When he was asked by the interviewer to expand on this, he seemed to have little justification beyond a preference for round figures. The grab disturbed me profoundly, because it was becoming clear to me that the Australian environment was having great difficulty sustaining its current population, let alone a rapidly growing one.

Ever since Paul and Anne Ehrlich published their landmark book *The Population Bomb* in 1968, the issue of population has been on the environmental agenda. But relating population size and growth to environmental degradation at the national level has proved to be an extremely complex and difficult task. Some economists thought they could dismiss the Ehrlichs' entire thesis (that population increase was a major environmental threat) simply by demonstrating the faultiness of some of the book's claims, but despite some errors in estimating resource size, the general argument is sound, for at the most fundamental level there is clearly a relationship between sheer numbers and the strain any species puts on its environment. When it comes to human beings, it's the relationship between population size, technology and level of consumption that determines the impact on the environment.

Because Australians are part of a global network of trade, that impact is disseminated globally. This makes it relatively easy to trace a direct relation between Australia's population size and the effect of such global phenomena as greenhouse emissions, but rather more difficult to estimate when it comes to local issues like salination. Still, we are beginning to learn the extent of the damage that we, as a nation, have inflicted on our soils and water as we have striven to maintain our standard of living and increase our population. To argue that the size of our population is irrelevant to such impacts is to ignore the fact that they result from the way we make our way in the world.

The thing that worried me most about the population debate in Australia in the 1980s was that absolutely no one in political power would acknowledge that the environment might be an issue. As we sang our national anthem, we all proclaimed that, "For those who come across the sea, we've bounteous plains to share." Yet no one was looking at the health of those plains or giving any thought to what their carrying capacity might be.

Whenever I tried to take up an environmental perspective on population at that time, I was shouted down as a secret racist. And almost every politician had an answer to the dilemma. It was inconceivable, they said, that technology and human ingenuity would not come to the rescue and "save" the environment.

It was during the Hawke era that immigration to Australia reached an all-time high. Had it continued at the rate set by Hawke in 1988, Australia would have reached a population of around 40 million by 2067. This may not seem like a large figure by world standards, but for a continent like Australia it is massive – remember that only 7 per cent of the land is classified as arable. At a conference on population held in Melbourne in mid-2002, author and researcher Dr Barney Foran calculated that Australians could live sustainably at our present population size if each of us reduced our demand for resources by 60 per cent. Put another way, this means that if we maintain our current level of affluence and resource demand, a population of around 8 million is all that is sustainable. Foran's calculation flies in the face of social reality and tends to produce knee-jerk reactions such as, "Well, who should go then?" It nevertheless neatly underlines the severity of the environmental problems we are struggling with. And it is worth recalling, in the light of Foran's findings, that if Australia had not experienced the massively increased immigration of the post-war period, and fertility had stayed at 1930s–early 1940s levels, Australia would have a population of 7.6 million today.

The CSIRO's landmark report on population, *Future Dilemmas* (to which Foran was a major contributor), was released in November 2002. It recommended that, taking into account Australia's environmental problems, the nation would be well served by a lower immigration intake, one that would deliver a population of 20 million by 2050. So incensed were the population boosters by this finding that the release of the publication was followed by an upsurge in the old assertion that population and environmental degradation are in no way related. The Federal Minister for Immigration, Philip Ruddock, for example, was reported in the *Australian* to have rejected any notion of a "carrying capacity" (an optimum population) for Australia. Making an analogy with sheep farming, he remarked, "If you add value to a sheep property, then you can carry more sheep." This assertion is undeniably true, and Ruddock's analogy between sheep and humans – in this respect at least – holds good. But the crucial question is whether we are in fact adding value to the national "property". Or are we, through Ruddock's immigration program, just increasing the stocking rate in a nation whose assets are being rapidly stripped?

The question of the relationship between population and its effect on the environment is one of the gravest facing us today, yet no other issue generates as many beautiful and not-so-beautiful lies. One of the most oft-quoted and most fervently believed-in dogmas (at least by the right) is the principle that human ingenuity can overcome all environmental problems. It is, I think, one of the most transparently false of all hopes. Let's look at the evidence: it is abundantly clear that right now our ingenuity is baffled and in disarray. It is losing the battle against the severe environmental problems that confront us as we try to maintain the most basic elements of life – our air, our soil, the quality of our water – and to conserve the biodiversity that is the continent's gift to its inhabitants.

As we have seen, global warming is a monster looming before us, and we lack even the slender leash of Kyoto to restrain it. If our much-vaunted ingenuity cannot get even the humble Kyoto Protocol signed according to a reasonable schedule, what hope do we have of acting in time for the full onslaught of global warming? A look at salinising soils reveals a similar countdown. Australians are clearing native vegetation faster than ever, our efforts at increasing water-use efficiency are exacerbating the problem, and not all the human ingenuity in the nation has been able to rescue the land now degraded with salt, much less that under imminent threat. There is, I repeat, no fix in sight. The fate of our rivers looks a little better, for here at least we have a sound plan – a widely acknowledged way towards remediation of the Murray – but even so no one as yet has acted to implement it, and time is of the essence. Consequently water quality continues to decline nationally. Although some endangered species have recovered (as a result of the efforts of groups like the Australian Wildlife Conservancy), the loss of biodiversity also continues apace, largely driven by land-clearing, river degradation, feral animals and the inability to manage fire. From the perspective of the early twenty-first century it looks as if, for all our ingenuity, the Australian farm is going to hell in a handcart.

The other argument often heard from the business community is that more people equal more dollars, which can then be used to fight environmental damage more effectively. While this sounds unassailably true, experience over the past few decades discloses a different reality. The wealth of Australians skyrocketed during the 1990s, yet our battle to save the environment continued to go dismally. No doubt it is the human condition that aspirations should be in the clouds while feet remain mired in mud. But only a person blinded by ideology could believe that a combination of human "ingenuity" and increased societal "affluence" is winning these battles. And who is so blinded by dogma that they would rapidly increase the nation's "stocking rate" at such a time?

Finally, we must acknowledge those idealists who believe that Australia can develop a "Singapore solution" to its problems, that it should simply abandon its hinterland and import its food, living in the manner of a clever country like Singapore, exporting high technology from a sun-bathed Aussie paradise. For one of the world's half-dozen or so reliable food exporters suddenly to become an importer of food itself poses certain difficulties, and of course Australia's degraded land will not simply look after itself in the event of irrelevance. Such utopian fantasies ignore the realities of a place like Singapore, for such entrepots can only arise in very special circumstances. The world has perhaps half-dozen of them. It also ignores the cruel fate allotted to Australian manufacturing over the last few decades. If, of course, we see signs of such a future emerging, we should all cheer and call for increased migration forthwith. But don't hold your breath ...

In the face of the irrefutable, there are those who still argue that human resourcefulness, armed with technology, will save us if only given enough time. This may be true, because no one knows the future, but it represents the kind of blind hope that encourages despair. To increase the nation's "stocking rate" on the mere hope that ingenuity will eventually triumph is like gambling the nation's assets at the Crown Casino. If and when we see signs of environmental recovery, is the time to think about increasing our population – anything else would be irresponsible. To do so beforehand is not only to put the well-being of future generations of Australians at risk, but to accelerate that risk. We should never entirely rule out the possibility of taking that risk, but we should only do it when the reasons for doing so, whether strategic or moral, are overwhelming.

In the light of Australia's dire environmental danger, it is reasonable to ask whether the nation can continue to sustain itself in the medium to long term. This is a question with massive political and human implications. In other words, is it possible that the needs of future generations of Australians are actually irreconcilable with those of refugee migrants in desperate need? Can we sustain a compassionate, caring nation on our difficult continent? Before such a question can be answered, we need to examine the relationship between human rights and the environment.

A respect for the intrinsic dignity of all human beings dictates that, above and beyond everything else, population and environmental policies must obtain the best possible outcome for the greatest number of people. In the context of Australia's environmental predicament, this includes non-Australians facing desperate situations overseas, generations of Australians still unborn and, of course, the citizens of Australia, the voting public. Far from there being a conflict between human welfare and the environment, the dilemma, seen in this way, becomes an indissoluble unity. Degrading environments inevitably lead to degraded people, now or in the future. It is only if we take too narrow a view, intent only on the present, that there appears to be a conflict.

Let us summarise Australia's position and then draw an analogy that neither minimises the opportunities nor underestimates the burden that a humanist (indeed a humane) perspective enjoins on us. Australia, as we have seen, is labouring under the burden of a profound environmental crisis for which there is no solution in sight. It is a middle-sized power and a middle-sized economy, quite unable to save the entire world or indeed anything but the tiniest fraction of the most distressed part of it. Finally, there are indications that, at current levels of consumerism and technological capacity, the continent has too many people – and still we continue with an immigration program and shy away from a population policy to regulate it.

The analogy of "lifeboat Australia" is not heard very much these days, but in the light of these facts it is still apposite. Imagine the captain of a sunken vessel. He has not gone down with his ship but instead has clambered onto a lifeboat that still has some space for others, at least on a calm sea. He floats in the midst of drowning souls. Which way does he turn? To the nearest person, or to the largest group? To go towards either means leaving many others to perish. And what does he do, after the decision has been made, when some desperate soul attempts to clamber in and those already aboard cry out "Full!" This, in essence, is the burden of human justice when it must be enacted in a world of limited resources. It's a world known not only to stricken sea captains but to doctors in overburdened hospitals, generals at war, even politicians. All of them know that to achieve the greatest good one must sometimes stand by helplessly in the face of almost intolerable suffering. Can we afford to take the present number of refugees in terms of our present resources? No. Can we afford not to take the refugees in terms of our Judaeo-Christian inheritance and the humanist philosophy that has come out of it? No.

So what degree of responsibility do we have? It is to one particularly significant element of Australia's immigration program that we must now turn. Where do those refugees come from who currently make up one twentieth of our immigration intake, and why have they fled? We must face up to the fact that Australia's foreign policy has created considerable misery in the world. By the time this essay appears, Australians may be fighting in the Middle East, in a war that is not sanctioned by the United Nations. More desperate refugees will doubtless spill out of the arena of conflict. How have we prepared to receive them? Will they be welcomed here? The questions only sound rhetorical because our record is so poor.

One would have hoped that we would have learned from the past. The tragic voyages of the Vietnamese boat-people were just one of the many awful consequences of the Vietnam War. It was an unjust war that any nation whose sights were set firmly on a humane and equitable outcome would never have engaged in. Instead Australia went "all the way with LBJ" and a decade later we were shocked by the tide of misery that washed up on our shore. Not only shocked, but at times openly hostile. Nor was Australia blameless in the Indonesian take-over of Papua in 1961 and the invasion of East Timor in 1975. Or, more recently, in places like Afghanistan and Iraq, and in our acquiescence in the bloody mess that is Palestine today.

I do not believe for a moment that Australia can set the world to rights. But I do believe that the genuine long-term national interests of Australia can only be protected through a respect for the intrinsic dignity of all human beings. This is because, while dictatorships and governments come and go, the people will always be there, and it is with the people of the world (however hideous the governments they endure or flee may be) – with our true neighbours, our allies and our competitors – that we must co-exist forever. So our immigration program must go forward carrying the burden of this sorry history and the moral obligations that go with it, though the form this moral obligation takes should not in a simple way be pre-determined. The greatest tragedy, however, would be to continue with a foreign policy based on the same narrow, short-term self-interest that has so palpably failed us.

In this area of grave moral responsibility, we need to act on the basis of deep consideration, flexibly and with a view to achieving the most humane end, avoiding the satisfactions of both false prejudice and false piety. Australia's foreign aid budget is so small that it hardly bears scrutiny, and what is worse, it is never tallied with or made to work in complement with the refugee element of the migration program. How do we know that the small amount we spend on refugees and foreign aid is producing the best outcome in terms of reducing human misery? Could we spend our foreign aid more strategically to minimise the misery that results in refugees? And is it possible that we should we spend more on foreign aid and less on refugees? Such questions are complex, and human lives can hang in the balance as a consequence of the choices we make. There is no comfort in any of this, but whether we are humanists or Christians, Jews or hippies, if we honour the debt we owe to those we share the earth with, we will realise the complexity of what is at stake. Sometimes it will mean that we must gamble on whether a political activist will languish in jail or a child die of cholera in a pool of filth. The Australians making such decisions need to get wisdom and with it moral courage and understanding equal to that demanded of the imaginary captain of our sunken ship. In other words we need true leadership in this area, a leadership that is equal to the hard choices to be made, because without it Australia will go forward into a miserable world that is in part a world of its own making, ever less able to make a difference.

Now, however, it's time to turn to those in the lifeboat shouting "Full!" as well as to a few of those who cry "We've room to share!"

Before settling upon some solution, Australians need to acknowledge their own natures. Not the noble vision we have of ourselves as the nation of the "fair go", but the rough beast that raises its head whenever we forsake our commitment to a world view based on humanism or kindness, that kinship that links us with the stranger.

The Hawke years brought a significant shift in immigration policy. While the majority of migrants still hailed from Europe, for the first time since the gold rush significant numbers of Chinese were also coming. Their visibility on the Australian social radar was heightened unforgettably by the Prime Minister's tearful announcement that 40,000 visas would be issued to the Chinese students already in Australia at the time of the Tiananmen Square massacre. It was a compassionate gesture that did something to ameliorate the horror felt by many ordinary Australians who had watched events in Beijing unfold nightly on their television screens, and as a gesture it was spontaneously welcomed by many people. Nevertheless, no one could argue that it was the best way of expending Australia's limited capacity for population growth in the cause of human rights, nor the most astute political move in a deeply xenophobic nation.

As Hawke gave way to Paul Keating and levels of migration remained high, strains began to show in the Australian social fabric. The nation was, famously, to "become part of Asia". It was an announcement easily misunderstood, for the speaker was also the prime minister who felt that Australia was positioned at the arse end of the world and who patently preferred the sophistication of Europe.

In poll after poll an overwhelming majority of Australians – around 70 per cent at times – stated that they believed the level of immigration was too high. Many otherwise reasonable Australians believed that their way of life was under threat by large numbers of people whom they thought of as different from themselves. Not a beautiful lie, perhaps, but in an insular, relatively homogeneous society, an understandable one. Many educated Australians were deeply shocked and dismayed by the veer to the right that Australian politics took in the 1990s, yet it seems almost impossible that they did not see it coming. Pauline Hanson was elected to Federal Parliament and went on to become a potent force in Australian politics, and partly in response to the same wave of feeling John Howard dramatically lowered immigration. The Labor Party was routed federally and remains in disarray today, with immigration in its various manifestations the great and ongoing stumbling block to its recovery.

The re-election of the Howard government has seen a different set of concerns arise in relation to immigration. The Chinese and other recent Asian migrants have long since been accepted by the majority of Australians. Now it's the Muslims who "really are different, you know" – who represent a threat in the public mind and supposedly will never fit in. Following the *Tampa* crisis, a rigorously applied policy of detention has seen potential migrants who arrive unofficially be trapped within an inhuman bureaucracy. Implementing this terrible system may well have choked off the arrival of boat people, but it has disgraced our nation in the eyes of the world and materially damaged our commitment to the fair go. And still the Howard government has no population policy. In fact, Philip Ruddock has stated on television that a national debate on population would not result in a population policy. It is my worst fear that the development of a policy has been held up because in a policy vacuum a government can do many things to serve sectional interests at the expense of the national good.

The self-interest of various sectors of Australian society is powerful indeed, and all too often they justify their lobbying with more retailing of beautiful lies. Business needs population growth to generate huge and easy profits, and it is from the business community that some of the most rosy-spectacled optimists about population hail. In the United States, illegal immigration is openly welcomed by business, for it provides a pool of unprotected workers that can be used on a "just in time" basis, like any other commodity of modern business enterprise. Australian industry is not yet calling for more illegal immigration, but its cries for more migration overall are incessant and clamorous. The boosters justify their position either by denying that the environment represents any constraint at all on population growth, or by asserting that changing technologies and human cunning will solve all problems the future may throw up.

It is possible, I suppose, that the majority of captains of industry see their interests and those of Australia as one and the same, especially when it comes to population. Why shouldn't the free marketeer believe in the free flow of people and also subscribe to an ideology of freedom for all? Still it is remarkable to hear CEO after CEO, Chairman of the Board after Chairman of the Board, call for more migrants just as they are cutting their own workforces – something we have seen happen again and again at great human cost over the last few decades.

I believe that the self-interest of the business immigration program is morally dubious on a number of grounds. First of all, it can serve to undercut Australia's commitment to its own education programs. By bringing in trained graduates from overseas we both deprive developing nations of desperately needed expertise and undercut our obligation to train our own youth. If you take the narrow view, it makes "business" sense to get someone else to pay for the education of your workers, but it is liable to be deeply damaging to Australian society. Secondly, as American employers all too clearly know, immigration can undercut the conditions enjoyed by workers. When the growing crisis in Australian nursing became evident last year, the knee-jerk reaction of some was to call for increased immigration of nurses. This conveniently ignored the fact that to work in the creaking public health-care system was becoming an increasingly insupportable option for many dedicated health-care professionals, which was why they were leaving in droves. To bring outsiders in to a failing system is clearly not a way to pool resources or to proffer a solution to a problem that can never be solved by minimising expenses.

It looks to me that immigration is to Australian business what Snowy water is to the irrigators – an essentially "free" good that has resulted in addiction to an environmentally unsustainable business model. By world standards, Australia has one of the highest rates of immigration anywhere on the earth. In the absence of a population policy, this should at the very least be a matter for lively discussion between all Australians. It should not be swept under the carpet, nor should the implications for the future fail to be spelled out.

Environmentalists tend to take a precautionary approach in population matters: that is, to allow for increased immigration when the gains from improved technologies, or a commitment to more moderate affluence, are clearly in place. To paraphrase Philip Ruddock, let's increase Australia's stocking rate only when we see substantial improvements in the quality of the farm. Environmentalists are often criticised because their approach threatens to leave many desperate people beyond the reach of our compassion and assistance. Yet no conceivable refugee program could allow Australia to assist all or even proportionately very many of the desperate people of the world. Instead we need to create a better world by wielding our foreign aid budget and foreign policies as weapons that aim squarely at delivering the greatest benefits for the greatest number of people we know we can touch, however indirectly. I believe that as things stand today, it is important to limit Australia's immigration intake. I do not know at what level that intake should be, but I think that establishing this figure should be the first job of those entrusted with developing a population policy, and that the development of such a policy is crucial to the responsible government of Australia.

It is also fair to ask what the composition of the overall immigration program should be if it is regulated under the umbrella of a just and humane population policy. Again the question is difficult to answer, but one determining factor should be the relative cost-benefit of spending dollars overseas as opposed to spending them on assisting refugees to settle in Australia. Another would be the effect of events beyond our control.

Wars and revolutions, famines and acts of God, might mean that Australia would have to give refuge to large numbers in one year, fewer in the next. This would need to be offset against other elements of the program, or perhaps by "borrowing" against intakes for future years.

Would illegal immigrants still be locked away under inhumane conditions under such a program? Possibly, but not necessarily. It all depends upon us, and how successful we are in recognising, then changing, the beast of xenophobia that raises its head in Australia whenever we lose sight of those ideals – let's call them humanist as a shorthand – that constitute our better nature.

As we feel our way towards socially just population and environmental policies, we will make such progress as we can armed only with an imperfect knowledge of key issues and needful constraints. Would a population of 30 million, at current levels of affluence and technology, irretrievably cripple the environment to be inherited by future generations of Australians? Will technology radically change so that a larger population can be accommodated? Or will Australians be content to settle with less affluence so that there can be a larger number of us? These are some of the great unknowns. But we cannot just go on as if these questions do not exist. The fact that we cannot escape our uncertainty will mean that mistakes will be made – and this may lead to further environmental or social damage – but a brave leadership true to a properly humane vision deserves to be forgiven such mistakes. After all, what is the better option?

Australia today is a sweet-and-sour nation. Many rural Australians are making a difficult and fundamental transition in human ecology aimed at achieving environmental sustainability. That will require – in the short term at least – a sacrifice of affluence for many of us. But this is the sweet side of things, for it lays the foundations of a truly Australian nation and a genuine end to colonialism. Yet even as Australians are doing this, we are spreading misery all around us, and the sour taste that misery engenders will come back to haunt future generations of Australians as surely as the bitterness of salty soils and rivers.

It was a visiting American, Samuel Langhorne Clemens, known to literature as Mark Twain, who said that Australian history reads like the most beautiful lies. I think that Clemens felt that way because the histories he was given to read were indeed filled with romantic falsehood. From now on – for the next little while at least – the history we create must be more mundane. It should tell the story of a small country that did the best it possibly could for the people and the environment of the world.

By way of a coda, I offer this manifesto for creating a better Australia.

1. Resolve that the human consideration – creating the greatest good for the greatest number – must underlie all of our environmental, immigration and foreign policy decisions.

2. Appoint a federal minister responsible for the welfare of non-Australians. This person should be charged with creating the greatest human benefit by the use of our combined immigration and foreign aid budgets.

3. Calculate the cost to future Australians of all initiatives that affect the environment. This should include matters related to population. Such costs should be explicitly addressed in all environmental and immigration decisions.

4. Focus our environmental effort as strenuously as possible on the key issues of maintaining healthy water, air, soil and biodiversity. These issues need to be fought as we would fight a war, for the future of Australia is at stake.

5. Implement programs that empower individuals to fight, on their own terms, the threats of global warming, salination and biodiversity loss. Such programs are now urgently needed in the cities of Australia.

6. Deliver, through our education system, a clear program that does full justice to the humanistic and Enlightenment heritage which has been ours since the time of the First Fleet. Young Australians need to discover how these principles have underpinned our nation.

7. Maintain Australia as an outward-looking, confident part of the global network of humanity. We should foster understanding rather than fear, engagement rather than isolationism, as the most fundamental requirement of our democracy.

8. Try, at every step, to expose the lie of *terra nullius* and so move towards a post-colonial Australia that is truly at home with its environment and history, where the Aboriginal and the Asian and the White Australian can believe in the truth of a history, and the justness of a future, that is so much more than a beautiful lie.

SOURCES

Essay sources and occasional supplementary material are given below. Page numbers indicate where the quotes etc. appear.

2   On population figures without post-war immigration see A.J. Marshall, *Australia Limited*, Sydney, Angus & Robertson, 1942, p. 105.

5   See http://www.australiaday.com.au/address.html for Rick Farley's 2003 Australia Day Address.

6   Charles Darwin, *The Voyage of the Beagle* (1839), Middlesex, Penguin, 1989, p. 326 & p. 332.

7   *Pauline Hanson's One Nation*, Manly, One Nation, 1998, p. 131ff.

7–8   For Buckley's account of cannibalism see Tim Flannery ed., *The Life and Adventures of William Buckley*, Melbourne, Text, 2002, p. 197.

8–9   L.R. Hiatt, "Ignorance and a question of shame" (letter to the editor), *Sydney Morning Herald*, 26 April 1997.

11   On Dawes' accomplishments see A. Currer-Jones, *William Dawes, R.M.*, Torquay, W.H. Smith & Son, 1930. p. 7, quoting from a lecture read to the Royal Australian Historical Society by Professor G. Arnold Wood on 28 August 1923.

14   Tench's account of the ending of the 60,000-year separation can be found in Tim Flannery ed., *1788*, Melbourne, Text, 1996. p. 41.

19   On these mammal extinctions see Dorothy Tunbridge, *The Story of the Flinders Ranges Mammals*, NSW, Kangaroo Press, 1991.

20   On role of fire in mammal extinctions see Tim Flannery, *The Future Eaters: An Ecological History of the Australasian Lands and People*, Chatswood, Reed, 1994, pp. 237–241.

23   Figures of proportion of immigrants from "A Survey of Migration," *The Economist*, 2 November 2002.

24   Account of festival of Husain is found in the *Sydney Gazette*, 26 March 1806; also in Tim Flannery ed., *The Birth of Sydney*, Melbourne, Text, 1999, pp. 200–202.

25   Peter Cunningham, *Two Years in New South Wales*, London, Henry Colburn, 1827; also found in Tim Flannery ed., *The Birth of Sydney*, Melbourne, Text, 1999, pp. 240–241.

28   On early acceptance of Chinese diggers see Eric Rolls, *Sojourners*, St Lucia, University of Queensland Press, 1992; Hotham quote from *Sojourners*, p. 127.

36   Marc Raisner, *Cadillac Desert: The American West and Its Disappearing Water*, New York, Viking, 1986.

36   Playford quote from Ticky Fullerton, *Watershed*, ABC Books, 2001.

38   Rachel Carson, *Silent Spring*, Boston, Houghton Mifflin, 1962.

38   A.J. Marshall ed., *The Great Extermination: A Guide to Anglo-Australian Cupidity, Wickedness and Waste*, London, Heinemann, 1966.

48   See the Wentworth Group's *Blueprint for A Living Continent*, WWF Australia, 2002.

50   Moo-cow Mitchell quote from Hugo Beckle, "The salinity crisis: looking back and forward," *Prospects for Biodiversity and Rivers in Salinising Landscapes: Conference Abstracts*, Albany, WA, 20–27 October 2002.

50   W.E. Wood, "Increase of salt in soil and streams following the destruction of native vegetation," paper presented to Royal Society of WA.

51   On threat to over 850 unique plant species and lack of broad-scale solution see Hugo Beckle, "The salinity crisis: looking back and forward," *Prospects for Biodiversity and Rivers in Salinising Landscapes: Conference Abstracts*, Albany, WA, 20–27 October 2002.

51   D.L. Nielsen, M.A. Brock, G.N. Rees & D.S. Baldwin, "The effect of salinity on freshwater systems," *Prospects for Biodiversity and Rivers in Salinising Landscapes: Conference Abstracts*, Albany, WA, 20–27 October 2002.

51   On fifty-fold increase in salinity of wheatbelt rivers see S.A. Halse & J.K. Ruprecht, "Salinity in south-west rivers and the likely effect on aquatic biodiversity," in *Prospects for Biodiversity and Rivers in Salinising Landscapes: Conference Abstracts*, Albany, WA, 20–27 October 2002.

52–3 On computer modelling of river system see Kevin Goss, "Environmental flows and salinity management: competing demands for water," *Prospects for Biodiversity and Rivers in Salinising Landscapes: Conference Abstracts*, Albany, WA, 20–27 October 2002.

58   On Ruddock and "carrying capacity" see Paul Kelly "Deep Green Dilemma," *The Australian*, 9 November 2002, p. 28.

65–6 On US business attitudes to illegal immigration see "The Longest Journey," *The Economist*, 2 November 2002.

Graham Richardson

If ever there was an essay which needed to be written it is Amanda Lohrey's *Groundswell: The Rise of the Greens.*

Now that the Greens are enjoying electoral success, some examination of their past and some musings about their future seem to be in order.

When I first became an official of the Labor Party at the end of 1971, it was generally accepted that only 10 per cent of the electorate was not tied to a major party. This 10 per cent were the swinging voters of the day. Even the name "swingers" indicated that voting for a major party was the only real choice.

The only minor party of any note at that time was the Democratic Labor Party, and by then its death throes were well and truly under way. There was no room for small parties because, overwhelmingly, voters were tied either to Labor or to the Coalition. The Democrats had not yet kicked off and the Greens were not yet any kind of political force.

In just three decades, the percentage of voters committed to one or other of the major parties has really plummeted. Even being kind to them, 75 per cent would be a maximum figure. It is that decline in base support, particularly on the Labor side, which has given real impetus to the rise of the Greens.

Disillusionment in the major parties has been growing for all of these last thirty years. The euphoria of Labor voters when Gough Whitlam was elected didn't last long. The harsh realities of government proved too great a hurdle for a Labor Party which had spent so many long and bitter years in opposition. Too many of the men who became ministers in that government were too old, too stupid and too embittered from their long walk in the wilderness.

Whitlam had a vision. He could dream the big dreams but he didn't really know how to manage those around him. Within three years his government had descended into utter chaos. Kerr's intervention is probably the main reason that the enormous number of Australians who worshipped Gough (let alone the man himself, who never suffered from a lack of self-esteem) have been able to erase completely from their memories the turmoil of ministers getting tied up with a spiv like Khemlani, a notoriously distracted treasurer, sackings, comings and goings, a faltering economy, an overnight 25 per cent cut in tariffs – just to name a few.

This litany is not to be taken as any attempt to deny the wonderful, magical policy movements in health, education, heritage and the infusion of huge sums of money into helping out the far-flung neglected suburbs of our major cities. It is meant, though, to remind all those Whitlam worshippers that today's icon was massacred in the polls in 1975 and again in 1977 by an electorate deeply disturbed by his years in power. Many who had voted Labor for the first time vowed never to do it again.

Malcolm Fraser did the same thing for the Coalition. Given a massive majority and for a time a majority in both houses, Fraser could achieve almost nothing. The notorious tax indexation promise, dredged up to win an election and jettisoned just as quickly when its cost proved too great for the budget to bear, set the tone.

The economic mess of 1982 (when John Howard was treasurer) was the last will and testament of the worst prime minister since World War II.

Conservatives had seen too much incompetence and too many broken promises. The Coalition's electoral base was fractured and more swinging voters, or at least voters prepared to look for other alternatives, had been created.

During the Fraser years, the birth of the Democrats with their slogan "Keep the Bastards Honest" created a destination for at least some of these displaced major party supporters.

The '80s and '90s should be described in terms of economic rationalism rather than Labor dominance. It was the strict adherence of both sides of politics to this ideal that provided the framework for policy. Any minister who wanted to oppose or even temper this grand plan was by definition "irrational". If you thought the pace of tariff reform was too fast, you were a troglodyte. More importantly, if you dared to challenge the orthodoxy of extremely tight budgets, then you were a real moron.

That kind of climate didn't breed much Cabinet debate. It didn't guarantee over time that Labor's base would continue to stick blindly to the voting pattern established over eons. The base continued to fracture, as did that of the conservatives.

Since Whitlam's investment in education, more and more of the sons and daughters of the working class received higher education. They bought homes and wanted to send their kids to private schools. They worried much more about their mortgages than about third world poverty. They refused to follow the lead of their parents in voting Labor.

The seat of Macquarie in far western Sydney is a classic example of this phenomenon. A decade ago this was a safe Labor seat. Now the Liberals' Jackie Kelly appears to have an unassailable lead. It is mortgage debt country with a capital "M". If you drive around this seat it is really hard to understand why it isn't still a safe Labor seat. John Howard probably has a better understanding of these people than most.

Many of them are very new Liberal voters. They are not yet welded into the base, they haven't yet secured their permanent political home. The party best able to represent their interests will always be in with a show.

The disaffection of Labor voters, as demonstrated by the difficulty (at the moment almost the impossibility) of keeping the first preference vote at or over 40 per cent, has in recent times come mainly from the party's working class base. But it must always be remembered that Labor's base is made up of a significant minority (probably somewhere up to a quarter of the ALP vote) of professional, academic and better paid voters.

Their critics call them "chardonnay socialists". They are said to be a self-proclaimed elite who know what is better for the country even if no one outside their group believes it. Nothing any critic might say, though, can alter their legitimacy as a part of Labor's constituency.

The one thing that has turned Australian politics on its head more than any other in my lifetime has been refugees, and that issue is epitomised and immortalised by just one word – *Tampa*. Like most of us I wish I had never heard of this god-forsaken boat. Up until its arrival Kim Beazley was looking good. Sure, his lead in the polls had been whittled away over the months of June, July and August, but he still had, I believe, enough of a lead to win an election against John Howard.

That lead evaporated in an instant. I have never seen the mood of a nation switch so swiftly. A big majority of the electorate had been unhappy with "boat people" for a long, long time. Pauline Hanson for all her intellectual inadequacies had shown how the dispossessed, unhappy older white Anglo-Saxon Aussies reacted to an opportunity to express their resentment at the funds directed to Aborigines or refugees. For a few years she ploughed the fields of latent racism with unerring accuracy.

The more educated part of the Labor's base just couldn't cop Beazley (or indeed Crean) going along with putting refugees behind barbed wire. Men, women and children were to be incarcerated all over the Pacific and yet another fragmentation of Labor's base occurred. Whatever lefties remaining in the ALP jumped ship and headed for the Greens, whose stance on the issue was unequivocal.

As an aside: to be successful in any form of endeavour you need luck. With all due respect to the undoubted political skills of John Howard, how lucky was he when the *Tampa* came along? Sure he handled it brilliantly and the election a few months later turned into a personal triumph for him. But how much would history have been altered if the nearest ship to a sinking Indonesian vessel had been from the US, the UK, France, Germany or any other European country except Norway? The one country Australia could afford to insult or ignore was a country with only three million people and with which we have almost no trade. Would we have sent our armed forces to board an American or British ship? – I don't think so.

The alienation of Labor voters, either blue collar or professional, from the party's base had been going on for three decades, and it meant that the Greens had a larger group of disaffected voters whose affection was up for grabs.

Up until the *Tampa*, One Nation had been a focal point for disaffected Labor and conservative voters. John Howard swallowed up that latter lot in one gulp. But just as voting Liberal was the natural result after the destruction of One Nation, the *Tampa* left some of Labor's constituency with nowhere to go but the Greens.

At the same time, as Lohrey correctly points out, the Democrats had set about destroying themselves. Lohrey says you can't put the disintegration of the Democrats down to Natasha – and she's right. No, you can put it down to Meg Lees. Trying to hold down a group like the Democrats was never going to be an easy job. They never did stand for anything in particular, and the day that Lees put her hand up for the GST the betrayal felt by the Democrats voter was palpable. An army of Natashas couldn't have kept them together after that. Add to that problem the child-like, puerile antics of Andrew Murray, the pathetic ineptitude of putting Brian Greig in the leadership and then the lacklustre performance of the latest leader – whatever his name is – and you have a recipe for death and disaster.

Obviously, many Democrats deserters will have headed straight for the Greens. With all the decay in Labor's base in particular, and to some extent in the conservative base as well, the Greens should be doing well.

You don't need to read too much into their successes either. Lohrey says that, "The message from the Western Australian state election of February 2001 was this: after a period of stagnation in the '90s the environment was back on the political agenda." Well, the scandal-ridden Court government had been ripe for the picking for some time. The Greens did well in that election but Labor's performance was not too shabby either.

Amanda Lohrey's essay was written before the Victorian state election, but here the dangers of predicting too much for the Greens was underlined. Despite confidence in their capacity to poll well enough to win a couple of inner-city electorates, the Greens couldn't quite make it. After the Cunningham experience, perhaps they thought the kind of results achieved there could easily be replicated elsewhere. Life is rarely so simple.

In Cunningham, a Labor member had resigned years before the scheduled election and Labor had not given its traditional base too much to cheer about in a very long time. In Victoria the combination of a very popular Labor leader with enough pro-environment policies being put out by the government meant that the Greens could do very well but still fail to gain any lower house seats.

I suspect that a similar result in New South Wales is almost certain. Labor, under Bob Carr, will have a fairly comfortable majority. If the Greens were going to have a win in the lower house, it would probably be in the seat of Port Jackson – an inner-city seat where a council run by a coalition of greens, assorted leftists and old Laborites has managed to oppose pretty well all development. That is an attractive attitude to voters looking for not much change in their expensive, privileged backyards. Nonetheless, my betting would be that Labor's Sandra Nori will hang on narrowly and the Greens will do well without winning.

It is an old irony that Bob Carr has created an impressive number of national parks and is accused by the usual potpourri of four-wheel drive and horse groups of locking up too much of the state.

At the same time many Greens claim Carr just doesn't do enough. On election night Labor's victory will very much be put down to the leader, just as it was, quite correctly, in Victoria. Leadership does matter in politics, and here Lohrey rightly identifies the position occupied by Bob Brown.

Brown has absolutely no charisma. Still that does not seem to matter in a country that managed to make John Howard its prime minister. Brown is definitely no Winston Churchill when it comes to oratory. Still, that didn't stop Howard either, did it? He does have one quality that few if any politicians have. He is totally credible. He never says anything he doesn't believe. He never seeks to compromise. He doesn't twist his words to avoid giving a commitment or to duck a difficult question. Brown conveys honesty and conviction. He is an absolutist.

In 1987–88, the Hawke Cabinet debated the whole issue of World Heritage in Tasmania. The debate lasted weeks, taking up three full cabinet meetings. Cabinet very narrowly approved a big extra chunk of Tasmania being put into World Heritage. After the decision, on his own authority, Bob Hawke put even more of Tasmania into World Heritage than his Cabinet had approved. When I told Bob Brown of this great victory, which had only been achieved with much of my blood being spilled on the Cabinet room floor, all he could do was express his great disappointment that the discussion did not go far enough.

He can drive you mad but you always know exactly where he stands. There are many Australians, and they are not all Greens, who have yearned for someone to be like this in Canberra. Brown's popularity will be central to maintaining the Greens' forward movement.

The problem is that these days he speaks far more about refugees and wars. To build this new Green constituency he must stray far and wide from an environmental agenda. Then he and his colleagues must find the way to clasp all these new people, drawn not by saving trees but by the pursuit of broader social objectives, to their Green bosoms. I wish them all the luck in the world – I fear they are going to need it.

Graham Richardson

Jack Waterford

I'm with Amanda on the view that a green vote, loosely described, is here to stay, and that the nature of politics has changed enormously over the past twenty years. But I think that the causes are more complex and more general than she describes, and the risks greater.

The change in politics may have been catalysed by environmental consciousness, and may have been, in hard times, sustained by it. But the consciousness that now makes up the broader movement is, it seems to me, far wider and, as a consequence perhaps, less deep. Therein lies some of the risks for the party, such as it is.

The shift in politics is more symbolised than caused by the bipartisan spirit, between Labor and Liberal, on economic rationalism. It also reflects the changes brought about by our becoming a post-industrial society, one where traditional conflicts between capital and labour seem old hat, but new ones, particularly affecting the environment or personal amenity, seem more pressing. Where much of the agenda of an old generation of reformers (particularly on human rights) has been ticked off and, even if it didn't produce the nirvana we had expected, there do not seem to be great reforming causes any more. And where a new selfishness and lack of community focus in the middle (or, as John Howard would put it, "aspirant" or "battler") classes has seen a collapse of institutions, a cynicism about organised human activity and a view of a world as hostile and uncontrollable once one ventures beyond the front gate. And where, most immediate human needs being more or less satisfied, people have become increasingly suspicious of those who want to appropriate our wealth to others. Where parties are not much more than brand names, marketed by professionals, and where consultation or grass-roots involvement in policy-making is a complete farce. Where politicians, themselves increasingly remote from community organisations, consult pollsters on "what they are thinking out there" and believe the evidence supports a culture of meanness and exclusion, a sense of helplessness against major economic and social forces, and "choice".

Perhaps they, or some of them, are thinking like that out there, but there are also deep spiritual hungers, a feeling that there must be more to life than acquiring a DVD or doing another house extension, an uneasiness about the atomisation of the family and the evidence that we have been mining, rather than farming, our heritage, and some recognition that even the battlers are doing well by comparison with marginalised groups such as Aborigines. People are still suckers for moral leadership, for ideas that lift them out of the mire and show them the light on the hill, and for leaders and politicians who have a sense of purpose, some charisma and some nobility.

Such leaders, post-Whitlam, are in short supply from conventional politics all around the world. They are hardly evident in the conventional institutions either. When they do emerge, it tends to be from small-scale issues that in some manner encapsulate some of the community unease and thus make them major figures. As Amanda says, the phenomenon of a Bob Brown has some parallels with the phenomenon of a Pauline Hanson: at first instance, moreover, it is the capacity to articulate discontent which is probably more effective than the promise that one has a vision of a community in which such discontents would disappear.

The focus of the Greens is still deeply environmental, and so are its primary activists, but the base has widened to embrace a full range of ideas, not least after the 2001 election when Labor betrayed its membership base on refugees, whether on the argument of being a small target or because some actually agreed with the Howard government. The moral disgust felt by many traditional Labor voters is profound and unlikely to be retrieved in the short term, certainly not by gestures such as those made by Carmen Lawrence.

Yet if the "grass-roots" nature of the green movement has nurtured and developed the party, and new styles of meetings and consensus politics have kept people involved and committed, it is quite possible to overstate the "democratic" nature of such movements. In fact they are far more susceptible to small groups of activists, not least to those with agendas almost calculated to drive away recruits. The grass roots flourish only when there is a campaign in action, and one cannot do it all of the time. Populist movements always have trouble with nutters (ask Pauline Hanson), but there are rich pickings to be found among some of the environmental fundamentalists, not least among those most focused on issues such as animal rights, militant vegetarianism and modern secular Manichaeanism. Likewise, the tendency of members of failed or stalled religions

– Stalinism, Trotskyism, postmodernism and feminism, for example – to seek to harness and turn vibrant movements to their own purposes is a continual worry, the more so when such zealots tend to have the time to attend, and wait out, interminable meetings. There are many more Roundheads than Cavaliers, Methodists than Catholics, sin-banners than tolerators, and anarchists than entrants. Even within the environmental movement ideological divisions – between, say, the "realists" and the fundamentalists – run deep and tend to be resolved more by the charisma or decisiveness of leaders than by exhaustively reached consensus. As the movement becomes broader, however, the contradictions will abound, and in ways that will prove hard to reconcile.

The history of co-operation between Greens and Aborigines, for example, is not good: every time they have jumped into bed together, one or the other has complained, in the end, of rape. There is a natural sympathy with Aboriginal aspirations, not least with land rights and the notion, rather fanciful, that Aborigines enjoy a deep spiritual relationship with their land and are natural nurturers of it. But the experience and the mutual suspicion is such that most smart Greens will utter no more than polite slogans of support, eschewing much involvement. They cannot avoid some reconciliation forever.

There are fundamental contradictions between members of coercive tendency and those whose focus is on gentler persuasion, likely in time to create problems in constituencies such as the rural religious. The Greens, and Bob Brown personally, won much moral credit over their forthright stand on refugees, but had been rather more marked, some time before, for being opposed to immigration at all. That's not necessarily a contradiction, but it can underline some of the issues of focus. Aggravating the problem, of course, is the fact that many activists, of their nature, differ from broader members, yet have disproportionate influence. Leaders – and the Greens have had some excellent ones – find that coalitions of the willing (or available) are not the best company.

Amanda deals only in passing with the separated Greens brethren in Western Australia. Politicians from other parties found Greens Senators from that state (Wendy and Tinkerbell as they came to be known) impossible to deal with because they were politicians of an entirely different stripe, entirely unwilling to negotiate, compromise, to trade support on one issue for another or to make up lists of concessions or achievements they wanted as the price of reluctant support. They came to be regarded as anarchists almost determined to prove that the system could not work. By contrast, Bob Brown commands a deep respect, even from his political foes, without ever being accused of being a "realist" (dread sin) or a trader, and there is as much respect for his political nous as for his principles. It cannot be said that the contrast with the Western Australian Greens was simply because they were of a "new politics" that old politicians simply could not understand.

In many of these respects, the Greens are far better off than the Democrats, now almost certainly in terminal decline. As Amanda notes, the Democrats originally pitched themselves as being in the centre – to the left on social issues, more conservative on economic ones. But they never had a real constituency, let alone one to which they could be accountable, and the fiasco of their party government is something Greens should study carefully. Just as significantly, once the Democrats decided to become "participators", even arbitrators between positions taken by the rival big parties, they were squeezed: themselves bastards and unable to go as far as the Greens on almost any issue of principle, particularly the environment.

A time will come, however, when the Greens must actually participate in government, as they do in Europe. That's the test of capacity to compromise, though it is different from the role the Democrats invented for themselves, or the disastrous affair in Tasmania when the Greens supported (as the rope supports the hanging man) the Labor Party.

Jack Waterford

Greg Barns & John Cherry

This year, Bob Brown celebrates the twentieth anniversary of his election to the Tasmanian Parliament. Yet, according to Amanda Lohrey and some political commentators who should know better, the Greens are Australia's newest political fad and are destined to replace the Democrats. We would caution against such bland assertions, as history tells us that politics can change. The Greens have been up before and then failed to live up to the media's expectations. In any event, there is no rule that says it has to be an either/or choice – Democrats or Greens!

The Australian Greens party was formed in 1992 as a result of a merger involving state-based Greens parties. At the 1993 federal election, Bob Brown unsuccessfully contested a lower house seat in Tasmania, gaining about 14 per cent of the vote, although a second Greens senator was elected from Western Australia. The next three years saw the Greens rapidly expand, returning MPs for the first time in the WA, NSW and ACT elections. During 1995, the Greens polled higher than or within half a per cent of the Democrats in 17 of the 22 AC Nielsen polls and 16 of the 28 Morgan polls. They collected over 24 per cent in federal by-elections in Wentworth and Kooyong. In short, they were on a roll.

Yet, when it came to the 1996 election, the Democrats outpolled the Greens 10.8 per cent to 3.3 per cent nationally, and in every state except Tasmania. It would be another six years before the Greens recovered to be in a position to challenge the Democrats again.

What went wrong for the Greens in 1996 and are there lessons for 2003?

Simply put, the question is more correctly what went right for the Democrats. When it comes to federal elections, voters continue to differentiate between state and federal issues. The Senate, as the House of Review, is widely supported in its role, and accordingly tactical voting – voting for a major party in the House of Representatives and a minor party in the Senate – is common. The Democrats, as the specialist balance of power party, are respected as the party best placed to hold the balance of power and ensure that common sense, fairness and accountability govern the Senate's decision making rather than ideology and political grandstanding.

Over the last ten elections, the Senate vote for the Democrats has been on average 2.6 per cent higher than the vote in the House, as recorded in public opinion polls.

The Democrats also tend to do better in federal elections (where their Senators command a national profile) than in state elections. Over the past decade, the Democrats Senate vote has been, on average, 3.9 per cent higher than the vote in the nearest state election, while the Greens vote has been 2.2 per cent lower.

Can the Democrats recover? History suggests that they can: they have done so after a botched leadership change in 1990, the Kernot defection in 1997 and the instability/difficulties leading up to the GST-induced leadership change of 2001. This does not preclude the Greens also being successful, although Bob Brown will know from his twenty years of experience how fickle political fortune can be.

It should also be noted that the Greens voter and the Democrats voter are not necessarily one and the same. The former is a protest party of the left, the latter a party of the centre-left. The Greens' platform of opposition to globalisation, which enshrines the ability of workers to strike in almost any situation (thus it was able to attract some hard-left unionists in the recent Victorian state election), as well as its capacity to package itself as the perennial naive player on the political scene despite the fact that it is in reality part of the "furniture", means that it attracts left-leaning ALP voters – those who want to protest against the "system" – as well as those for whom the forestry debate is front and centre of their political activism.

The Democrats, on the other hand, attract voters concerned about sharing the benefits of globalisation equitably, maintaining the primacy of liberty in a democracy and ensuring that human rights informs policy prescriptions and outcomes. These voters are found across the political spectrum.

In fact, in the current national debate on liberty versus security, and in relation to other issues such as government policy on refugees, the need for symbolic as well as practical reconciliation with indigenous Australians and the growing power of executive government, the pull of the Democrats remains.

Greg Barns & John Cherry

Brian Coman

In the most recent issue of *Quarterly Essay*, Amanda Lohrey has given us a very good account of the Greens in Australia and why she thinks that their fortunes will continue to rise. No doubt the performance of the Greens in the recent Victorian elections and their earlier win in the federal seat of Cunningham can only increase her own confidence and the confidence of those who agree with her general thesis.

And yet, when one digs under the surface of this rather rosy picture of the Greens and attempts to analyse the basis of green ideology, certain antinomies are apparent. Indeed, Lohrey herself is not wholly unaware of these and, in the very last sentence of her essay, she admits that the chief difficulty for the Greens in the future will be "in keeping up with the complexity of expansion within the ecological constituency and in maintaining a balance of forces within their own movement". Those opposing "forces" she speaks of have to do with the very nature of our understanding of the term ecology and, in the final analysis, come down to opposing philosophical ideas which are as old as our civilisation itself. That they can be "balanced" or reconciled in some way is, to say the least, questionable.

Indeed, the modern version of this ancient "human-in-nature" debate shows even less sign of drawing towards some resolution. There is now an entire philosophy journal devoted to the subject of "environmental ethics", and debates employing hugely intricate argument upon subjects such as "intrinsic value in nature" still rage in its pages.

One can, I think, divide modern ecological thinkers into two broad and opposing groups – resource ecologists and radical ecologists. These two terms correspond reasonably well to the shallow ecology/deep ecology split proposed by the Norwegian philosopher Arne Naess in 1973. Resource ecologists constitute the mainstream group since most governments espouse the basic tenets of their ideas and they enjoy widespread public support. Put simply, resource ecology views the natural world in an anthropocentric manner, but seeks to place constraints on human use of natural resources. It sees humans as having certain obligations towards the natural world but those obligations are essentially to secure the present and future well-being of humans, both in terms of the maintenance or improvement of aesthetic values as well as the more basic, instrumental requirements of natural resources (food and water, clean air etc.). These obligations are most clearly manifested in the various national and international policies now in existence – maintenance of bio-diversity, sustainable development, pollution controls, world heritage listings, endangered species legislation, and so on. By and large, resource ecology has its basis in a purely material and anthropocentric understanding of the universe and it is not concerned with any metaphysical understanding of the natural realm. In this, it differs markedly from radical ecology.

Radical ecologists completely reject the notion of a human-centred cosmos and call for fundamental changes in the way in which humans view their place in the natural order. Drawing heavily on evolutionary theory, they see humans as no more than intelligent apes whose activity in nature since Palaeolithic times has been such as to "unhinge" them from the rest of the natural order in a way that is potentially disastrous, not only to themselves as a species but to the whole of the living world in general. The transformation of nature through human work is a wholly negative development, separating humans from the rest of nature. They see the root cause of this separation as being a fatal dualism in which humans have set themselves apart from and above the natural order. The solution to this problem, they suppose, is to change completely the way in which we view our place within the natural order such that we become eco-centric rather than anthropocentric. We are no longer to suppose that we enjoy any position of eminence or superiority as a species save for our intelligence, which demands of us that we regard the remainder of the natural order in the same way that we regard our own species. We then re-enter the ecological web of life in a fully harmonious way, operating with the same general principles as the rest of the created order. The notion of human work in nature fits very awk-wardly into such a schema.

Inherent in radical ecology is some notion of spirituality, albeit often of a rather feeble nature. Some radical ecologists think of a form of "world soul" or *anima mundi*, but then construe this in terms of a sort of web of interactions between all organisms and, indeed, between organic, living things and the inorganic. This notion goes back to Ernst Haeckel (1834–1919), a German biol-ogist and Monist philosopher who used the term *oekologie* to describe the web that linked organisms and their surrounding environment. Indeed, it is from his writings that the modern word "ecology" is derived. Humans then become simply "a node in the vast, interconnecting web of life". Others go much further and, drawing heavily on their own interpretations of ancient mythologies, speak in terms of "eco-spirituality" or "sacred landscapes". It is significant that few, if any, of these constructions of spirituality go beyond a form of degraded pantheism, and even fewer attempt to understand the human-in-nature question from within the traditional, metaphysical understandings contained in the major world religions. Part of the reluctance of radical ecologists to embrace a fully transcendental approach in this area stems from a perceived difficulty in aligning this view of nature with that presented by the popular, Darwinist approach. Transcendental beliefs are seen as giving to humans an undeserved degree of ascendancy or, indeed, a realm of existence which lies outside that allowed by the science of ecology.

Views from both resource ecology and radical ecology are found in green politics. They are often mixed together in a very strange amalgam and this, I think, is evidence that the Greens as a political party have not really come to grips with the basic ideology they purport to represent. Indeed, Lohrey's own position vis-à-vis the "human-in-nature" question is not easy to discern and she seems to reflect those basic contradictions that I believe to be inherent in all green ideology. She appears to adopt the general posture of the resource ecologist in her description of green ideology, yet her language betrays a certain ambivalence. On page 81 of the essay she supposes that, "No matter what our position is within the political spectrum we are most of us programmed to accept some plundering of nature as the price of progress …" The word "plundering" is significant. So are the words "programmed" and "progress". For someone who believes that the Greens are about "sustainable management of resources", use of the word "plundering" is inappropriate. In the West, since the time of Plato and up until no more than fifty years ago, the idea that human use of nature constituted "plundering" would have been thought very strange indeed. It is a word much favoured by radical ecologists. Other species "dominate" or "modify" environments, but humans plunder. I do not suggest that Amanda Lohrey is such a radical, merely that she has taken up certain of their notions by some form of unconscious osmosis. We are all influenced in this way. After several decades of viewing television nature shows, the deprecation of Western humanity becomes an acquired characteristic. The Lamarckian might be right after all!

Again, the idea that Western civilisation has been "programmed" to accept the plundering of nature is a notion characteristic of radical ecology. The "programming" that radical ecologists have in mind is the joint influence of Greek rationalism and Judaeo-Christian ideas, those very ideas that have formed the basis of Western civilisation. Yet, curiously enough, Lohrey believes in "progress" – a concept entirely characteristic of the modern West. As Professor J.B. Bury pointed out in his classic work *The Idea of Progress* (1920), the idea of progress we recognise today did not really come into being until the seventeenth century when Christianity began to lose its commanding position in the public life of the West. Some purpose and direction to history was certainly implied in pre-Enlightenment Christianity but the progress such Christianity conceived of was spiritual and eschatological, not material. It was the Enlightenment *philosophes* who gave us the modern notion of material progress (although it was certainly prefigured in the writings of Francis Bacon). With the Great War of 1914–18, the idea of progress lost much of its force, and events since that time have only served to diminish it further.

Today, the idea of progress is in very poor health, and if radical ecologists had their way I feel sure they would support some form of assisted euthanasia. Even resource ecologists are uneasy about the idea of material progress. True enough, certain American conservatives, such as Robert Nisbet, still believe in the idea of progress and some can even envisage an end-point where the whole world enjoys a liberal democracy of the American sort. Mind you, Nisbet thinks that progress cannot be achieved without a spiritual renewal. I wish him luck. A spiritualised version of Ronald McDonald or Colonel Sanders is difficult to imagine.

Put simply, the problem for the Greens is to accommodate the widely differing philosophies of their adherents while at the same time maintaining some identifiable and credible platform of basic ideas. The danger is that in reaching some highest common factor they will find that the resultant platform is so highly attenuated as to be next to worthless. Consensual arrangements, whether in terms of Christian ecumenism or ecological solidarity, tend to obey the second law of thermodynamics – entropy increases.

Finally, there is the question of social and economic climate. Anyone who has bothered to look into the history of ecology and green politics quickly realises that the "green" phenomenon had its birth in those countries with a large, educated, and prosperous middle class possessed of a strong liberal and Protestant culture (Britain, Germany and North America). Ecological concerns tend to arise when other, more basic concerns have been satisfied. Indeed, certain cynics have suggested that, having acquired the second car, the Jacuzzi spa and the speedboat, your average Western Joe now desires the next item on the list of "must have" acquisitions – a nice clean environment in which to enjoy all these things. All jokes aside, ecological concerns always work best on a full stomach, and if Australia's social and economic position deteriorates, then the popularity of the Greens may well subside.

Brian Coman

William J. Lines

Criticism is part of our culture. From Socrates and Christ through Rousseau and Marx to Arne Naess and Wendell Berry, half the business of Western culture has been self-criticism. We live not a single culture but a debating ground, not a monolith but a fertile confused corroboree of contending sources. People choose from an endless series of opposites – in philosophy, religion, politics, art, morals and manners – in living their lives and pursuing their truths.

Accordingly, the discontent driving those behind the Australian Greens Party and enabling them to articulate their frustrations lies not in the '60s, as Amanda Lohrey writes in *Groundswell* (pp. 4–5, 8) but is inherent in our culture.

I make this point not to censure Lohrey's perspicuous essay but to expand the discussion she generously makes possible. Her intelligence and range provide an opportunity to examine the Greens in contexts different from those permitted by the constrained level of political discourse normally prevailing in Australia.

I also write to balance a couple of biases. Only baby boomers believe everything began in the '60s. The history of human thought bestows little distinction on the period. Those years were special only because they coincided with the coming-of-age of a uniquely privileged, educated, historically ignorant and large cohort of youth.

The Greens inherited much older traditions. They are under the influence of cultural themes – primitivism, utopianism and emancipation – that predate the 1960s by five hundred years or more. The questioning these topics generate did not begin in the '60s and will not end with the Greens.

Lohrey's journalistic fluency should not blind us to another *Groundswell* bias: its Tasmanian slant. Part of the problem here is the lack of a clear distinction between the Australian Greens and the conservation movement. They are not the same. Both the Greens and The Wilderness Society (TWS) may be largely Tasmanian in origin (p. 16) but they are not identical and neither defines, delimits or represents conservation in Australia.

The Tasmanian bias discounts earlier conservation developments. Lohrey (p. 13), following other writers, claims the United Tasmania Group (UTG) was the world's first Green party. This bit of folklore requires context: there were precedents. In February 1971 three members of the Colong Committee – including Milo Dunphy – stood as Australia Party candidates in the NSW state elections. Their platforms were specifically conservationist. In September 1971 opponents of the proposed Clutha development contested NSW local government elections. Journalist Max Walsh viewed these developments seriously enough in the 21 September *Financial Review* to dismiss the conservationists as "eco-nuts". The formation and campaign of the UTG in the following year, with Milo Dunphy's help, was thus part of an Australia-wide stirring of eco-nuts. It was neither autonomous nor peculiar to Tasmania.

Also untrue is the claim (p. 16) that the 1982–3 battle for the Franklin River was "the first time a well-organised group of environmental activists mounted a successful propaganda war in the media and carried public opinion on a national scale". A decade before, a national and international conservation campaign saved the Great Barrier Reef. That struggle began in the 1960s following the formation of the Wildlife Preservation Society of Queensland (WPSQ), which sparked a new conflict in Australia: conservation versus development.

Indeed, the founders and leaders of WPSQ (including Judith Wright, Kathleen McArthur and Brian Fleay) plainly understood the real distinctiveness of modern times: the scale and extent of our deliberate manipulation and destruction of life compounded and driven by an unprecedentedly large and growing human population.

The Greens (formed in 1992, thirty years after WPSQ) are a relatively late, frequently incoherent and habitually unoriginal response to that situation. Not only do they encapsulate thoughts and behaviours inherited from much earlier times but also many party activists, educated during the 1970s and 1980s, give traditional cultural themes an orthodox leftist twist. Much party thinking runs in terms of class, gender, race, ethnicity and rights. These superstitions make it virtually impossible to talk intelligently about life on this continent. Understanding requires thinking in terms of flows, cycles, connections, exchanges and populations. Typically, people on the left are intellectually ill equipped to offer intelligent analysis of the human assault on the planet.

Leftist handicaps render the Greens incapable of tackling population, a matter dealt with forthrightly by the Wildlife Preservation Society of Queensland and other conservation groups contemporary with WPSQ. The Greens' 29-page manifesto, *Care for the Earth*, devotes just six paragraphs and six dot points to population. It is an appallingly inept declaration, a self-condemning jumble of clichés and qualifiers obviously written by a university graduate or, more likely and worse, by a committee of university graduates. The rest of *Care for the Earth* displays the same clumsy proclamation of dubious and incompatible platitudes.

Most Greens concede that the nearly 20 million people in Australia are living unsustainably. Yet the party endorses large and unlimited increases in population. This is not ecological accountability but hypocrisy arising from conflicting social justice and ecological aims. Bob Brown has consistently opposed controls on immigration, takes a strong pro-refugee stance, and avoids debating population issues. His silence is tantamount to censorship.

Many Greens do not view the party's principles as contradictory. Under the influence of the utopianism they inherited as part of their Western legacy, they believe everything they believe to be worthwhile and good is not only ultimately connected and compatible but actually mutually reinforcing. This foolishness reflects the credulity and self-deception that distinguishes our political culture. Those who support *Care for the Earth* never met a moral dilemma they believed could not be disposed of through wishful thinking. But while the principles of ecology may have changed some thinking (Lohrey, p. 81), they have not overturned the rules of logic or banished the need for coherence and consistency. On all these counts the Greens fail the test.

Even the Greens' many mystifications and inconsistencies are unexceptional. Confusion and contradiction are legacies of our civilisation, part of the excess of ideas bewitching and bewildering our fearsomely ambiguous and morally fraught world.

Care for the Earth belies Lohrey's claim (p. 82) that the Greens are the "bearers" of a "new gospel". On the contrary, it reveals the Greens as primarily humanist, therefore conventional. They put humans first. The manifesto's opening principle states: "protection and enhancement of the Earth's life support systems are fundamental to maximising human welfare". Framing conservation in terms of human interests echoes the ALP's 1973 party platform in which a "high quality" environment was seen as a "fundamental human right". This portentous but fatuous rhetoric of rights and entitlements undermines the obligation to care implied in the manifesto's title. But then *Care for the Earth* is not really about caring for the planet but rather specifies how people should gain from actions taken to mitigate ecological damage – caused by our attempts to advantage humans and extend human rights.

We live in an age of blueprints and solutions; we are futurist to the point of bigotry. The Greens reinforce these contemporary prejudices. Their manifesto offers yet another set of plans and blueprints: a bureaucratic recitation of "goals", "principles" and "targets".

Moreover, Green politicians use and endorse the modern jargon – assets, resources, life support systems, environment, management – that reduces the living world to an instrument for increasing human welfare. Their language perpetuates the outlook that underpins destruction and corresponds to established formulae, not new thinking.

Lohrey's "new progressive constituency" (p. 68) is a figment. There is no single Greens constituency demanding both "greater social and ecological accounting". Two interests lie behind the Greens and they conflict – as Lohrey acknowledges (p. 72). One is mainstream humanist besotted by leftist pedantry about social justice.

"Maximising human welfare", however, is not the chief concern of conservationists who have a broader, more radical and more generous view of protecting all life. For these conservationists, Earth comes first. They seek to save nature from humans, not for humans. Hackneyed notions of uplifting humanity do not sway them from the task of conservation.

Most conservation activists view the Greens as an adjunct to their repertoire of campaign tactics. The Greens are simply another means to advance the cause. They are not the only means or necessarily preferred. Sometimes they even hinder conservation. Conservationists well understand the points of conflict. TWS by-laws, for example, prohibit TWS campaigners, spokespersons, and members of management teams from belonging to political parties, including, most pertinently, the Greens. The separation is clear.

The Australia movement for conservation is a response by people to life under threat. It arises from a sense of place, is an expression of identity with this continent, and reflects concern to counter our assault on it. The conservation impulse is essentially nationalistic and has no intrinsic connection to the central party planks – human rights and internationalism – of the Australian Greens.

Kathleen MacArthur, for instance, linked Australian wildflowers and an indigenous Australian culture. In 1959 she wrote: "Our [culture] is still young and delicate and we must save the heritage of the land to nurture it." Judith Wright also believed in national heritage and advocated Australians identify with country. She thought that care and love of the land were essential in any effort to protect it. There is no mention of love in *Care for the Earth*.

The split between conservationists and internationalists in the Greens became apparent during the *Tampa* incident. Internal Green polling revealed that a majority of members and supporters, like most Australians, sympathised with the government's position. The party's humanist elite found this view unacceptable.

Dissent suggests many Greens supporters detected the same premise behind both sides of the subsequent debate: Australia is a big country with lots of space and must become bigger. Anti- and pro-refugee advocates, including prominent Greens, differed only in the means by which they were prepared to accept more people as immigrants. Otherwise, protagonists shared the assumption that Australia is unlimited, capable of absorbing any number of human beings. Dissenting Greens may have wished to repudiate the bigger is better outlook and figured that the government's position at least contained the potential for questioning population growth. The real principle here is that the state of the land – soil, air, water, vegetation, biodiversity – not our reception and treatment of refugees will determine the future of life on this continent.

Undoubtedly, the elite stand on refugees attracted new members and supporters, who in turn have likely turned the Greens into a majority pro-refugee party. But this stance contains dangers.

As the Greens focus more on social justice so they will more and more represent conventional politics and become increasingly irrelevant to what is truly novel about our times: a grossly excessive number of people demanding an extravagant living from a limited and degrading continent and a limited and degrading planet. As the Greens become more conformist and safe so they may attract still more disenchanted Labor voters. These voters are among the most intellectually rigid and doctrinaire in society. They support the Greens precisely because they recognise old promises, familiar rhetoric and reassuring abstractions.

But disillusioned Labor supporters cling to their socialist and quasi-socialist illusions, which they transfer to the Greens. They may well call for social accountability but they have little interest in or understanding of ecological integrity. And because their self-deceptions will never appeal to the majority of Australians, their support condemns the party to permanent marginal status. It may also lead to schism.

Many conservationists already talk about building a conservation party. I also know of rural-based activists joining the Greens hoping to counter what they see as excessive city influence. They are acutely aware that support from inner city ex-Labor voters weakens the Greens' commitment to conservation. Another campaigner described the Greens as "a raft for refugees from the Labor left". And Labor, particularly the Labor left, provides a poor and frequently antagonistic base for conservation.

The states are the main agents of destruction in Australia. Labor is just as zealous about conquering nature as Liberal or National – in many cases more fanatical. And at state level the increase in the Greens vote is small in comparison to the increase in the Labor vote, especially for governments pursuing aggressively pro-development agendas: Tasmania, Victoria, Queensland and Western Australia (where the Gallop government prepares to abandon its election promise of no old-growth logging). Unsurprisingly, Labor premiers often come out of the same highly educated doctrinaire left (Maoist and Trotskyist in two current instances) that now supports the Greens.

The point, again, is that the Greens are not the conservation movement and do not represent the conservation movement. Between the two lies considerable tension and disagreement. Support for the Greens goes up and down. Their electoral prospects wax and wane. The cause of conservation continues regardless. It does not depend on the fortunes of the Greens. It is much bigger and more important than that. Conservation is about protecting life, not about getting the Greens elected.

Bob Brown leads an internationalist, social justice party primarily concerned with human rights, which receive priority in any conflict with the conservation of nature. Through their manifesto, language and disregard of population the Greens reveal themselves as educated, ideological and brainwashed, little different from the educated, ideological and brainwashed people making the decisions precipitating imminent ruin.

The Greens are a generic political consortium, a facsimile of the German Greens, rather than an authentically Australian response to the unique life of this continent.

William J. Lines

Ian Lowe

The beginning of a new century can be the nucleus around which a new way of thinking crystallises. The changes of the 1990s will shape the twenty-first century in the same way that the developments of the 1890s, leading to Federation and the emergence of the ALP, shaped the Australia of the twentieth century. I believe a new politics is emerging: a struggle between the Greens and conservative forces that cling to the old economistic view of the world.

Amanda Lohrey showed in QE8 that the Greens were displacing the Democrats as a third force in Australian politics. The vote in the 2002 Victorian election confirmed that view. There are some signs that we may be witnessing a more fundamental change, a major realignment of politics. The growth of green politics in the late twentieth century had a clear parallel in the emergence of the ALP a hundred years ago. The ALP, like similar parties in New Zealand and the UK, began as the political wing of the trade union movement. Unionists saw that they could not achieve their goals by industrial means because of the fundamental imbalance between capital and labour, so they entered the political arena. That radically reshaped politics into a struggle between Labour parties and anti-Labour coalitions. The old differences between Whigs and Tories, between Free Traders and Protectionists, between urban and rural capital, or (in Australia) between different states, became irrelevant to the fight against the common enemy of unionism or "socialism".

Recently, Green parties grew out of the environmental movement for the same reason that saw the ALP emerge a century earlier. Activists saw that modern politics was so dominated by economic issues that their goals could not be achieved by community action or pressure-group politics. The growing support for the Greens shows that the community, increasingly disillusioned with the short-term economic goals of the ALP and its conservative opponents, is prepared at least to cast a protest vote against both by voting for the Greens or even the improbable One Nation. Increasing numbers are now voting for the Greens in a positive sense, embracing their new and different approach.

The mass media has been slow to recognise the broad political agenda of the Greens. Until the publicity for Senator Bob Brown's resistance to the Howard government's racist approach to refugees, the Greens were usually depicted as a one-issue party fixated on saving forests. Even this issue has come out of the hinterland into the political mainstream, with the emergence of such previously unthinkable groups as Liberals for Forests in Western Australia. The preferences of Greens candidates have been crucial to the election of state ALP governments in recent years.

The Greens' agenda, however, always included economic and social issues. Four Green MPs backed a minority ALP government in Tasmania after the 1989 election, under the terms of an accord between the parties. As part of the accord, I was asked to help develop an alternative economic blueprint, dubbed the "greenprint", for Tasmania. At a series of public meetings in major centres a plan was drawn up for sustainable resource use. This was a clear change from the traditional Australian economic approach of destroying our natural resources for short-term economic gain. Some of the ideas in the greenprint have developed: mainland consumers now pay premium prices for clean produce from Tasmania. Unfortunately, the accord collapsed just as I handed the greenprint over to the Department of Premier and Cabinet. The incoming Liberal regime had no interest in such new ideas as using resources sustainably, so Tasmania returned to the traditional mix of market forces moderated by ad hoc government interventions.

Since then, it has become more obvious that the old approach is not working. Within Australia, the Australian Bureau of Statistics released in 2002 a report that drew together indicators of economic, social and environmental performance for the 1990s. It found that all the economic indicators were positive, that the social indicators were mixed with some serious negative trends, and that all but one of the selected environmental indicators were negative.

The first national report on the state of the environment concluded in 1996 that achieving Australia's declared goal of sustainability would require integrating ecological principles into all social and economic planning. Instead, governments at state and commonwealth level not only failed this test but increasingly abandoned the entire notion of planning in favour of a simplistic faith in market forces. The second national report in 2002 observed that most serious environmental problems had worsened because the pressures causing the problems had increased.

At the international level, the *Global Environmental Outlook 2000* United Nations report concluded that the present approach is not sustainable. Global environmental problems are getting worse, while the inequality between rich and poor is increasing. "Doing nothing is no longer an option," the *GEO* report said. Growing concern about global climate change led to the negotiation of the Kyoto Protocol, an agreement to curb emission of greenhouse gases, but the treaty has not entered into force at the time of writing. To its shame, the Australian government joined with the Bush regime to form an Axis of Environmental Irresponsibility, two rogue states supporting environmental terrorism by refusing to ratify the treaty. The final declaration of the Global Change science conference in Amsterdam in 2001 warned that many crucial parameters of the Earth's systems are now outside the range of previous experience since the emergence of humans as a species, and that our science is unable to predict the consequences of these changes. The most hopeful conclusion of the *Great Transition* report in 2002 (from the Stockholm Environment Institute in Boston) was that it is still possible to halt the momentum toward an unsustainable future, but it will require fundamental changes in technology, values and desired lifestyles. Even with these positive changes and concerted effort, the report warned, it will take many decades to bring human use of resources into balance with natural systems.

The evidence is clear: fundamental changes are needed to achieve a transition to a sustainable future. The political question is also clear. Can the ALP, the force for progressive change in Australian politics for the last hundred years, reform itself to retain that mantle? Or will the Greens become the new political force?

The ALP had the chance to reinvent itself in the early 1990s but chose instead the path of pragmatic opportunism. The last years of Bob Hawke's reign saw the ALP returned to office on Green preferences and responding with policy initiatives that could have led to significant reform. Hawke set up the Ecologically Sustainable Development process, in which nine sectoral groups developed recommendations for change in such areas as agriculture, manufacturing, transport, mining and forestry. Though served by a bureaucracy that was predominantly hostile, the working groups brought forward proposals that commanded broad support. After being diluted by the public service, the recommendations were brought together to form the basis of the National Strategy for Ecologically Sustainable Development, adopted in 1992 by the Council of Australian Governments. While not a radical document, it would have seen significant changes in resource use and environmental protection. Earlier that year, environment minister Ros Kelly had represented Australia at the Rio Earth Summit, and in that capacity had supported Agenda 21, the overall programme for reform, and the Framework Convention on Climate Change, a draft treaty to address the most serious global environmental problem. Within Australia, the government had developed the National Greenhouse Response Strategy, a co-ordinated package of measures to respond in a sensible way to climate change.

The Resource Assessment Commission was originally a pragmatic solution to defer the political problem of a proposed mine at Coronation Hill until after the 1990 election, but it developed a methodology that was a forerunner of the more recent Triple Bottom Line approach. The RAC argued that it was possible to assess the proposal in each of three areas: its economic benefits, its social costs and benefits, and its environmental risks. Weighing these to decide if the economic benefits justified the social and environmental costs is a political value judge-ment, the RAC said, so it should be made openly and transparently by the elected government. The Hawke Cabinet eventually agreed to refuse permission for the mine, mainly because Hawke believed that the social costs to local Indigenous people were not justified by the economic gains to others. The decision was the last straw for the pragmatic forces who organised behind Paul Keating to depose Hawke.

As the new prime minister, Keating showed the permanent mental scars of his years of working with Treasury, who had persuaded him to champion a goods and services tax in the mid-1980s, ten years before he opposed its proposal by John Hewson and fifteen years before John Howard inflicted it on Australia. Keating clearly did not forgive the RAC for its approach of forcing politicians to make a transparent decision that weighed up the economic, social and environ-mental issues; it was disbanded. The Ecologically Sustainable Development proposals gathered dust in Canberra pigeon-holes. The Australian delegation to the second conference of parties to the climate change treaty in Geneva advanced the notion of "differentiated responsibilities", a serious attempt to derail the entire process by claiming that Australia should be given a more generous target than other countries, while at home the consensus recommendations in the National Greenhouse Response Strategy suffered the same fate as the Ecologically Sustainable Development policies. In short, the green tinge that the ALP had developed under Hawke was systematically expunged by Keating, his eraser being a mixture of market economics and traditional ALP-Right opportunism.

That history suggests there is little chance of the ALP recovering the baton of progressive change. As Sylvia Lawson wrote in QE8, it "needs a recovery of imagination". In the late 1990s and the early years of the twenty-first century, the only sense in which the ALP has been consistently progressive is that it tends to support unions against employers. ALP politicians are more likely to support a republic than Liberals, more likely to be in favour of reconciliation with Indigenous Australians, more likely to support womens' rights and less likely to support racist immigration policies, but these are differences of degree rather than clear divisions. In similar terms, the ALP has called for Australia to ratify the Kyoto Agreement and use more renewable energy, but there seems little chance of it adopting a thoroughgoing commitment to sustainability.

One obvious impediment is political. Just as the Liberal Party is beholden to commercial interests that would be affected by fundamental change and so can be guaranteed to oppose reform, the ALP's union base includes employees who would also be affected. While changes would create new industries and new job opportunities, the new commercial interests do not yet exist to support and influence the Liberals – and the new union members are not yet putting pressure on the ALP to shape its policies.

A second problem is the political process itself. As Paul Kennedy argued in *Preparing for the Twenty-First Century*, the politicians who succeed in mainstream political parties are usually those who have refined to an art form the practice of avoiding any threat to powerful interest groups. They are therefore, he said, most unlikely to adopt unpopular measures now in the interests of future generations, as long as they can argue that experts are divided or that more research is needed. In the case of complex environmental issues, it is nearly always possible to argue that the experts are not of one mind and that more research is needed. So the new breed of pragmatic ALP politician is very unlikely to go out on a limb for significant change.

Fundamental change is usually driven by the ideologically committed. That was the case in the overthrow of Keynesian economics in the Australian bureaucracy and its replacement by an economic fundamentalism that makes Adam Smith look humane and enlightened. The process was driven by a small cadre of zealots working within the system. The same argument applies to the change needed in thinking about our interaction with natural systems. There is increasing evidence that we are not using the natural resources of Australia sustainably. Soil salinity, the state of the inland rivers, soil erosion, climate change, the decline of rural towns and the collapse of traditional resource industries are trends that cannot be ignored. In the face of these evident signs of serious problems, the Liberal–National Coalition explicitly tells people to trust their economic management, with occasional cosmetic gestures like the Natural Heritage Trust, condemned by its independent reviewers as largely a waste of public money. As the alternative government, the ALP dithers ineffectively, unwilling to risk unpopularity by arguing for fundamental change. The Greens, ideologically committed to a different relationship with the planet and seeing social and economic decisions as being subsidiary to that relationship, are proposing a response that is radical enough for them to be the voice of the twenty-first century. The Cunningham by-election is, like the 1989 Tasmanian state election, a harbinger of the new electoral politics of the twenty-first century.

That does not mean that Australia will move swiftly and surely into a new mode with the Greens as the main reform agent in national politics, any more than the rise of the Labor Party in the early years of last century was guaranteed. An obvious question to be resolved is whether the Greens are organisationally equipped to handle their growing responsibility. That raises a final parallel with the Labor Party. Originally the political wing of the trade union movement, it has broadened its base over the last hundred years to include two other discernible groups: leftist activists who see the ALP as the most likely vehicle for progressive reform, and pragmatic opportunists who see it as the most likely vehicle for their own personal advancement. While the same process is less advanced in the case of the Greens, its beginnings can be seen. In the case of the ALP, the tensions between the different groups have produced one major split and a more recent division into factions that often seem more interested in their internal rivalry than the fight against their common enemy. If the Greens are to become the voice of reform and spearhead the transition to a sustainable society, they will need to be better than the ALP has been at the task of developing a united front and acting as a cohesive organisation. The division in overseas Green parties between the fundamentalists and the pragmatic realists may well become apparent locally once the Greens have real political influence.

Ian Lowe

Paul Rogers

Who are the Greens and where do they come from? Amanda Lohrey is right to ask this question in the foreword to her essay *Groundswell: The Rise of the Greens*. Having resigned at the end of 2002 from the Queensland Greens after seven years, and with some experience of the dynamic forces within the Greens, my observations may be of interest to readers of *Quarterly Essay*.

The closer one is to the party political process, the more one can misunderstand perceptions in the apolitical community, although I do not necessarily accuse Lohrey of this. For me, at polling booths, a common rhetorical question from putative green voters was: "You're for the trees, right?" Yes, they usually get that bit right, and although Lohrey gives an accurate and enjoyable historical account of green politics in Australia to date, it would have been useful to explore, in more depth, at least some juxtaposing influences, not always harmonious, that represent the aggregation of forces – the green palimpsest in her words (and Doyle's) that constitutes the Greens in Australia.

Unlike the Greens in Germany where the "realos" (realists, pragmatists) and the "fundis" (left idealists, fundamentalists) are somewhat distinct and quarrelling forces like factions of the ALP, the Australian Greens have, until recent times, been able to sublimate any internal friction and prevent it from receiving public scrutiny and analysis. Lohrey's essay tends to reinforce this comfortable image, and although she hints at diverse organic origins, she does not explore some fundamental philosophical and doctrinal differences within green politics. Yes, "for the trees" no doubt, but what else?

To give the hint, Lohrey records that Drew Hutton, Queensland Greens, anticipates problems with maintaining a balance between the social justice and the deep ecology formations within the party. Lohrey on Hutton from page 72: "What worries me most ... are those people coming from what I would see as an ideological social justice position, ideological leftism."

Within a week of the September 11 attacks on the USA, the Queensland

Greens had issued a press release condemning the attacks but suggesting that America deserved this terror for past actions including intervention in Vietnam and Chile. The release was also posted on the party's website. It has since been removed. No similar press release was issued after the Bali bombings, as one would expect, but the implications of such a position on September 11 can be interpreted in many ways, few of which are flattering for the Greens. Anti-Americanism is no substitute for thoughtful foreign policy.

I was appalled at this insensitivity considering that many nations, including Australia, lost citizens in these attacks – and that 21 per cent of the 2,800 killed in the World Trade Center were not even born in the USA. Yes, analyses of the reasons for September 11 are required. Yes, international social inequality undoubtedly plays a role in fomenting support for such violent movements. That it was exclusively a result of America's Middle East and overall foreign policy is a typical, one-dimensional response from the left.

I don't think that moderates in Algeria, India, Indonesia, Philippines, Singapore, Chechnya-Russia and Pakistan, nor in France or Germany, where attacks have been foiled, would agree either. This is not just America's problem. The US is a convenient and obvious focus, but a fierce, global, anti-secular and theocratic fascism should not be appeased or justified in my view.

Nevertheless, it set the scene for an Australian Greens anti-war stance that eventually led to my resignation from the party in November 2002. The paucity of intellectual debate or alternative viewpoints on the US and Australian response in Afghanistan also surprised me, considering that the worldwide intellectual left is somewhat split on the issue.

Further, circulating widely in the Greens was the infamous conspiracy theory that the CIA and Mossad were responsible for September 11 and that it was all to do with oil supplies that would be enhanced with a pipeline built across Afghanistan. Now I understand that all political parties have their ratbags, but I began to see this as parallel to the One Nation conspiracy theory that the federal government was behind the Port Arthur massacre (along with the UN) in order to hasten the introduction of gun laws. I was having no part of it.

It seems to me that the Greens' policy direction and rhetoric on "peace" is the least rigorous, most left-captured and most contradictory of all their policy. The left seem to have lost much moral relevance in today's world. Many opposed NATO intervention in Bosnia despite the documented barbarism by ethnic cleansing forces, including the terrible rape and murder of women. After September 11, in Greens' internet mailing lists, few were prepared to support military intervention in Afghanistan even though the Taliban were enslaving and persecuting women and denying them education, protecting organised terrorist murderers, withholding food from their own people and persecuting and murdering Shia Hazaras in the thousands, resulting in many eventually ending up as refugees in Australia.

I note that Kerry Nettle, recently elected Greens senator for New South Wales, expressed solidarity with the women of Afghanistan in her first speech to the Senate. It seems they now have some chance of a better life. This is a serious moral conflict for women of the left. Too many, in my view, choose not to tackle this complex moral issue.

I am an opponent of George Bush on most grounds, including his policies on greenhouse emissions and genetic engineering of food, and his administration's general economic and conservative agenda. Yet in some circumstances, difficult as it may be to accept, there is a greater good to support. Afghanistan was one case. Iraq is probably another.

The German Greens have evolved to the point where the fundis, and I interpret this as the "hard left", are less influential on policy and electoral processes. Joschka Fischer, the current Greens defence minister in the German coalition, supported sending German peacekeeping troops to Bosnia and he also supported the coalition intervention in Afghanistan. Not bad for a former anarchist and roommate of Daniel Cohn-Bendit! Watching former anarchists in the Australian Greens describe themselves as "Ghandists" and from the "peace movement" elicits a smile, if not a smirk from me. On the other hand, people are entitled to evolutionary political change, as some Australian personalities have demonstrated.

The non-aligned peace movement has an important role to play in world affairs and I support it, but it seems to me that many of its members have to do some serious multi-dimensional thinking. Just-war theory is valid. "Peace" is not necessarily synonymous with, and a result of, "no war".

Lohrey may be correct in suggesting that a political force with the authority to alter drastically the balance of power in Australia is emerging, but let's not get too excited just yet. In my experience a reasonably solid 8 to 10 per cent serious green vote has existed in Australia for many years, split between the Democrats, the Greens and some Labor. Now, with the Greens doing well in the latest opinion polls and the Democrats losing relevance quickly, the opportunity exists to build strongly on that support for the Greens. However, it would be a mistake to assume that the green vote is making inroads into suburbia. It has been slightly swelled by refugee and war talk issues and the demise of the Democrats. Nothing more.

Alas, any further movement to the left and focus away from an environ-mental agenda will see a ceiling reached very quickly indeed. I doubt that many want the socialist left of the ALP in Greens' clothing. Make no mistake, the old left are actively courting the Greens. Having backed a lame horse in the past, they now want another go. If this tendency continues, many environmentalists will slowly move on to concentrate on the work they have always done and the Greens will be left with the political rump of the left. Not a pretty sight! The Labor Party may find that wooing the green vote in a determined and credible fashion could be very fruitful indeed, but one doubts they are on anything but a path to mediocrity.

In conclusion, allow me this observation and emotional indulgence from my deep ecology roots. The urban humanist left that make up much of the influen-tial Greens membership needs to get their arses wet. They need to be alone and up to their knees in snow on a high pass or at night in the depths of the rain-forest in a tropical storm to sense the limitations and the relevance, in the longer term, of an organism called *Homo sapiens*. Paul Ehrlich once reminded us that "nature bats last." James Lovelock would, no doubt, agree.

Paul Rogers

George Seddon

As an historical survey of the major green causes, Amanda Lohrey's essay is excellent and a useful summary of all the things the Greens have been against. What the essay is not is a critical analysis of the green movement, and that perhaps is because it *is* still a movement rather than a political philosophy. There are two common threads: the STOP response (flooding Lake Pedder to the Ningaloo Reef proposal), always linked to a specific issue and largely negative; together with a diffuse awareness that things are rotten in the state of Denmark and that we can't go on as we are – a feeling I share.

Two current issues in Western Australia are the campaign to "stop logging in old-growth forests" and the development proposals at Maud's Landing (Ningaloo). Both issues are complex. There are beyond question some significant biological values associated with old-growth forests, but the forests still have to be managed: if they are not, destructive wildfires are almost inevitable. The campaign should be that our forests, all of them, should be managed to take account of a full range of values, and not just timber production. We don't have more than a vague idea of how to do this yet, but we should be finding out, and this is what reasoned *policy* would look like.

The Maud's Landing proposal is equally complex. I had grave doubts about the proposal as first mooted, but it has been considerably refined. I am very sure that the status quo in the area is wholly indefensible and that there is an urgent need of a major infrastructure overhaul, but it is far from clear where the money is to come from. The Maud's Landing proposal has now had guarded approval from the Environmental Protection Agency and it is not out of the question that a beneficial outcome could be negotiated, i.e. one with more benefits than disbenefits. Blanket opposition and a failure to recognise the gross inadequacies of the status quo are not helpful.

Both of these issues are serious enough, but they pale into insignificance against the massive environmental problem facing Western Australia, and indeed most of Australia, and that is largely inappropriate and unsustainable land use. A real policy would not only recognise this, but search for the necessary changes, ones, however, that have massive economic and social implications as well as ecological ones. But perhaps this is too much to ask of any Australian political party.

George Seddon

Gareth Evans & Sidney Jones

The 1969 Act of Free Choice was fraudulent. The evidence is overwhelming. The United Nations-supervised referendum, in what was then Western New Guinea, on independence or integration with Indonesia, was preceded by widespread use of force and intimidation by the Indonesian army. The 1,022 Papuans hand-picked to represent the population at large were interned for months before the vote and warned of the consequences of not choosing integration. The UN team of sixteen people was far too small to monitor what was happening, even had it any interest in doing so. The vote itself took place through different regional meetings in which some two dozen people were asked to stand up and give their views, all of whom made virtually identical speeches about their commitment to the "Red and White" – the Indonesian flag; the others, watched by Indonesian military personnel, were then asked to stand up if they agreed, which they all did. The unanimous result looked so bad that even European diplomatic observers thought there should be a few no votes, just for the sake of legitimacy. This was not, by any stretch of the imagination, a valid act of self-determination.[1]

Should the whole matter, then, be re-opened? The answer is yes, but we have to be clear-eyed about what is ultimately achievable. Would the objective of re-opening the Act of Free Choice be assessing the historical truth, acknow-ledging fraud or holding another referendum? The first is easier to achieve than the second, and both are easier than the third. While an historical re-assessment should ideally be conducted with full transparency by an impartial commission under the auspices of the UN, the truth-seeking process could start without the UN, and indeed already has. In December 1999, Dutch Foreign Minister van Aartsen agreed on the part of the Dutch government to authorise an historical re-examination of the Act of Free Choice. The US, Australian, and Indonesian governments could and should do the same.

But even setting the historical record straight is more complex an exercise than may appear at first sight. An issue arises for a start about time frame. To understand the injustices that culminated in the Act of Free Choice, do you just look at the period July–August 1969 when the "consultation" took place, or do you go back to 1962 and examine the New York Agreement under which the UN authorised the transfer of administration of West New Guinea from the Netherlands to Indonesia? Or back further? The time frame very much determines at whose door responsibility should be laid for the outcome. Another issue will be the quality of the evidence obtainable. The voices of Papuans obviously need to be heard. Systematic documentation is critical, based on interviews with Papuans, Indonesians, journalists and diplomats on what took place from 1963 to 1969.

It certainly cannot be assumed that anything but an international commission will begin to get at the historical truth. Under the original draft of Papua's special autonomy law, before it was diluted by the Indonesian Parliament, there was a provision for a Truth and Reconciliation Commission that, among other things, would examine the human rights violations that took place around the Act of Free Choice. That provision has been weakened and changed to the point that it now says a Truth and Reconciliation Commission should be formed to clarify the historical record on Papua *in order to reinforce Indonesian unity*. It is no wonder that the Special Autonomy Law in its present shape is not widely welcomed in Papua.

If all these difficulties are overcome and the historical record is set straight in a credible fashion, the obvious question is what happens next. Should the UN formally acknowledge that the Act of Free Choice was manipulated, in the interests of Cold War politics? Yes. But where does that leave the Papuans? A serious truth-telling exercise within Papua and among governments that were actively involved in endorsing the Act of Free Choice would undoubtedly give a boost to the independence movement. But the Papuan leadership would also have to understand some harsh political realities.

The UN giveth and the UN taketh away. With the General Assembly having embraced the Act of Free Choice, at least to the extent of "taking note" of the Secretary-General's report on it, it is difficult to argue that anyone but the General Assembly could approve any move to void it. But it is almost impossible to conceive of that happening. While global politics have changed in many ways since 1969, the African and Asian nations who aligned with Indonesia in 1969 to accept the results of a deeply flawed process would almost certainly do the same again now. Multi-ethnic countries brought into being in the 1960s through the demise of colonialism, and with questionable colonial boundaries defining their nationhood, are not going to want to see touchy questions re-opened. Nations with majority Muslim populations who voted with Indonesia in the General

Assembly resolution in 1969 are likely to back Indonesia again more than three decades later.

One of the strongest arguments that will be heard against re-opening the Act of Free Choice will be the precedent it would set for opening other UN-supervised referendums to scrutiny. By what criteria, it will be asked, will the UN determine that one referendum should be re-examined and another not? Many Indonesian army officers, civilian officials and pro-Indonesian East Timorese maintain that the UN-supervised August 1999 ballot in East Timor was marred by irregularities. Those claims were dismissed at the time, and the Indonesian government accepted the results of the ballot. But if a mechanism for reviewing a decision were to be created, would that not open the way for any member state of the UN to petition the Decolonization Committee for review of any past act of self-determination? To ask the question is to anticipate the answer.

There is unlikely to be any Security Council support for a review. The Security Council, whose powers to act are effectively untrammelled – at least for matters it decides are capable of constituting threats to international peace and security – could conceivably authorise a course of action culminating in a new Act of Free Choice. But even governments committed to historical truth are not very likely to change their position that Indonesia's territorial integrity is now a given. On this kind of issue, the dynamics, and the arguments, would not be very different from those in the General Assembly.

So far as the United States is concerned, to the extent that it has an interest in Indonesia, it is currently to keep the government of the largest Islamic state in the world on side, or at least neutral, in its war on terrorism and potential war against Iraq. This is not greatly different from the *realpolitik* motive of the US in 1962 for supporting incorporation of then West New Guinea into Indonesia as a means of arresting the feared slide of Indonesia into the communist bloc.

No UN action on Papua is thinkable without the support of the Indonesian government. People tend to forget that, regardless of the force of the argument about the illegality of Indonesia's original incorporation of East Timor, and the extent of the concern about current bloodshed, there would have been no intervention in East Timor without the ultimate acquiescence of Indonesia – reluctant as that agreement was. It is simply impossible to see any support within the current government in Jakarta, or any foreseeable government, for a new referendum on independence or integration on which an international movement could gain a toehold.

Yes, the same was said about independence for East Timor, and independence eventually happened. But this was the product of a wholly unforeseeable course and combination of events – the 1997 economic crash, the collapse of the

Suharto government and the elevation of an extremely idiosyncratic president. In all Jose Ramos Horta's many conversations about East Timor with Gareth Evans in his years as Foreign Minister the focus was on how to achieve autonomy: independence was assumed, by both of us, to be quite out of reach. Maybe we will both be proved wrong again about Papua, but the chances of the future running as fortunately as it did for East Timor are slim indeed.

All these political obstacles should not stop a re-assessment of the historical record from going forward. But Papuans should be under no illusion that international support for a new act of self-determination, let alone independence, is just around the corner. What a serious move toward re-assessing 1969 might do, however, is at least generate some momentum for the achievement within a reasonable time of an autonomy package capable of winning acceptance among Papuans.

At the moment, even the word "autonomy" has come to take on some of the same negative connotations that it acquired in East Timor: despite some of the very real devolution of fiscal and political authority guaranteed in the 2001 autonomy law for Papua, many Papuans see "autonomy" as Jakarta's code word for "status quo". It did not help that the law passed by the Indonesian parliament so diluted the original draft prepared by Papuan leaders that it fell short of the aspirations of even the most conciliatory Papuans.[2]

As long as the manipulations of the past remain deliberately obscured, both by the international community and the Indonesian government, there is going to be a high degree of suspicion about Jakarta's intentions for Papua's future. Setting the record straight on 1969 by itself is not going to end those suspicions, but it might help non-Papuan Indonesians understand why resentment against the central government runs so deep. It also might provide the basis for re-assessing what the relationship between Jakarta and Jayapura should be, not just in terms of what percentage of revenues can be retained, but on less tangible issues such as what history gets taught in the state school system and how demands for justice can be addressed. Whether autonomy will find more favour with Papuans than it has now remains to be seen. But without greater honesty about the past, the chances will be very much lower.

Gareth Evans & Sidney Jones

1 The evidence on the conduct of the Act of Free Choice has been most fully collected in a 2000 doctoral dissertation by John Saltford for the University of Hull.
2 See *Indonesia: Resources and Conflict in Papua*, ICG Asia Report No. 39, 13 September 2002, available on www.crisisweb.org.

John Martinkus

Gareth Evans and Sidney Jones are right. The Act of Free Choice was fraudulent. But where does that leave the Papuan people and the Indonesian state today? It leaves the Papuan people with a burning sense of injustice over their treatment by the Indonesian state, their dispossession from lands through mining, logging and other economic activity, with little or no compensation; and with a settled resentment towards a foreign community that has long ignored their economic and cultural domination by a country they believe has no right to be there. The Indonesian state, on the other hand, is today in a position to bury the issue of independence for West Papua before it gathers any momentum in the international community. It is doing this quite successfully in a number of ways.

Killing Theys

There is no doubt that the murder of Papuan Presidium leader Theys Eluay in November 2001 by Kopassus (Indonesian Special Forces) members threw the West Papuan leadership into disarray. The Presidium Council, which advocated a peaceful program for democratic and consultative change in West Papua, has never really recovered from the loss of its popular leader and the obvious challenge to its authority. Since the killing, Presidium leaders Thom Beanal and Thaha al Hamid have received regular death threats and are never without bodyguards. Two other Presidium members, in Wamena and Fak Fak, have been killed in suspicious circumstances. The intimidation has worked to an extent, and the Presidium has had trouble regaining momentum and popular support.

Interestingly enough, the trial of the Kopassus officers and men responsible for the death of Theys has been received bitterly here in West Papua. Human rights workers deplore the fact that the junior Kopassus private who actually did the hands-on killing (he strangled Theys) looks likely to be the only one sentenced. They believe that those who ordered the killing should be on trial as well. But in an Indonesian military court like the one where the case is being heard in Surabaya, that is simply not going to happen.

Forming militia — Laskar Jihad

It has a depressingly familiar ring to it, but as a strategy it seems to work. Kopassus and the Indonesian military are now forming and training militia groups made up of Papuans and Laskar Jihad (a Muslim militant group) operatives who have moved from the Moluccas to West Papua. In Sorong and Fak Fak, the Laskar Jihad have been training for more than a year. In Wamena in the highlands, the leaked documents of last year relating to the formation of a militia have been proved true, and human rights workers in Jayapura say that the militia there are now armed with M-16s. In the transmigrant settlements near the town of Arso and along the PNG border, locals say Kopassus has been holding meetings and recruiting at least ten locals from every village for service with the militia and training with Laskar Jihad groups now in the border areas. A series of unsolved murders of Papuans has accompanied this activity. The payments to local Papuans, from the reports I have received, range from 100,000 to 200,000 rupiah ($A20 to $A40) — relatively cheap, and, as we are still seeing in the ad hoc human rights tribunal in Jakarta relating to crimes in East Timor, a great way to deflect responsibility for abuses away from the military. The existence of militia in East Timor is still being cited by former Indonesian Armed Forces chief General Wiranto as proof that the conflict there was between two sides — a civil war, in other words — and that the military was simply trying to keep the peace. If serious conflict were to break out in West Papua, the same excuse would be used: it would be presented as a conflict between the separatists (the Papuan people) and those who wanted to remain a part of Indonesia (the militia armed and trained and paid for by Kopassus). The military would simply be trying to keep the peace and not be responsible for the inevitable one-sided bloodshed.

Diplomatic pressure

The most striking example of Indonesia's campaign to bury the West Papua issue came in comments to the press during Prime Minister John Howard's February visit to Jakarta. Security Minister Susilo Bambang Yudhoyono said he raised the matter with Howard. "There is a problem that is hampering our bilateral relationship. That is political activity in Australia, either by NGOs, certain individuals or universities, which in any way provides the opportunity for activities that could disturb our sovereignty in Papua," he told the press.

President Megawati also commented, "I conveyed to the Prime Minister the possibility of creating a policy for people residing in Australia but who are continually trying to bad-mouth Indonesia, because their actions can cause damage to the relationship between Australia and Indonesia." The result of this was an agreement between Howard and Megawati to have ministerial-level talks on the issue.

This pressure is already working. Melbourne's Royal Melbourne Institute of Technology recently cancelled a planned February conference on the West Papua issue, with the Vice Chancellor stating it was not the role of educational institutes to threaten the sovereignty of other nations. Since when have institutions of higher education throughout the world not questioned the political status quo? Isn't that one of the functions of higher learning?

The jailing of Scottish-born, Tasmanian-based academic Lesley McCulloch in appalling conditions in Aceh for five months was a clear signal by the Indonesian government of how they will deal with those foreigners who persist with their investigations into Indonesia's conflicts. In the end McCulloch was charged with visa violations and released in February this year, but only after a series of farcical hearings made clear to those present that her major crime and the reason for her incarceration was talking to those who were suffering abuses at the hands of the Indonesian state. About the time she was released, Susilo Bambang Yudhoyono stated that the government had decided not to allow foreign-initiated proposals to conduct historical or political studies that could intervene in Indonesia's sovereignty. He made special mention of the provinces of Aceh and West Papua.

Intimidating those they can't buy off

On 28 December a group of gunmen fired forty rounds into a car carrying the wife and daughter of Johannes Bonai, the director of Papua's sole human rights organisation, ELS-HAM, as they travelled between the Papua and PNG border. His wife and daughter, along with another woman, were seriously wounded. No effective investigation has been carried out because when a team of Indonesian police returned to the site on 1 January, they too were shot at. A preliminary report by the police contains a statement from a witness who saw Indonesian soldiers emerging from the bushes after the shooting. Since then Johannes has received messages on his answering machine that are the recorded sounds of people being tortured. ELS-HAM was the first organisation here to blame the military, and Kopassus in particular, for the murder of Theys (which they have since admitted to in court) and the attack on the Freeport mine workers last year that killed two US citizens and one Indonesian. He believes the attack on his wife was just the next level of intimidation.

Last year seven Kopassus officers kidnapped a tourist guide who they believed was working with the OPM. I interviewed him here the other day. His back was a mess of deep knife cuts inflicted while he was held to the ground in Kopassus headquarters. The Kopassus officers wanted to know whom he worked for and said they were going to kill him. Luckily he escaped and is now in hiding, but the two weeks when he was beaten, slashed and given nothing but salt water to drink have left him incredibly weak. One of his friends who went to enquire about his condition with Kopassus has been kidnapped as well. He is probably either dead or being held and tortured at the moment in a building I can see from my hotel.

Activists here talk about how they are regularly offered money to cease their activities. A priest who works with activists told me that he was offered a very large sum of money to stop this work. Similarly the tourist guide believes he was identified to Kopassus by someone who was offered a large sum of money to inform. These are the ways Indonesian power is implemented at a grass-roots level.

Autonomy ... where now after the division of Papua?

In direct contravention of the autonomy program that governments such as Australia and the US have endorsed as the only way forward for Papua, Indonesian President Megawati has issued an instruction to proceed with the division of Papua into three provinces. A military governor has already been appointed to one of the regions, and despite continued protests it seems the division will go ahead. Even those Papuan officials, such as the current governor, Jaap Salossa, who endorsed the autonomy proposal, now oppose the division because they see it as a way for Jakarta to circumvent the current Papuan civilian leadership, re-impose military authority and avoid passing on to Papuans the financial benefits of autonomy. According to Presidium leader Thom Beanal, "What we see now is that Jakarta does not know how to resolve the West Papua problem. They haven't had a committed approach and therefore, being paranoid, they try to divide and colonise by bringing in more military. If they divide the province in three, they can bring in more military personnel. They will not only multiply the number of military personnel, they will carry out more operations. In terms of Papua being a priority for them, that is how they are going to handle it."

According to the OPM on the northern Papua–PNG border, last December was a time of intense military activity. They claim Indonesian troops pursued them deep into PNG territory and that at least two of their number were killed. They claim to have killed as many as fifty Indonesian soldiers in counter-ambushes. Whether that figure is wildly inflated or not, conflict has definitely resumed on the border. PNG Prime Minister Michael Somare's special adviser Stephen Pokawin told the *Australian* newspaper that the existence of Laskar Jihad, which he linked to Jemaah Islamiah, was a major security problem for PNG on the border, and that PNG feared incursions by this group. The statement merely showed that the Indonesian military on the border is continuing its programs of using Laskar Jihad and militia proxies to wipe out the OPM regardless of the human rights abuses this will inevitably entail.

These are the actions of a military that remains unpunished for its previous abuses and similar programs in East Timor. Unfortunately for current Australian policy on West Papua, a repeat of the diplomatic excuses and the covering up for the Indonesian state's methods of control will lead us down the same path of appeasement that we followed in East Timor from 1975 until 1999. The future for Papuans looks very grim indeed, which is why the issue won't go away no matter how much policy makers wish it would.

The contribution by Gareth Evans and Sidney Jones is a welcome one in that it recognises the central illegality of West Papua's incorporation into Indonesia. What is needed now is for international organisations to start paying attention to the situation in West Papua in spite of the concerted Indonesian efforts to keep foreigners physically out of West Papua and pressure organisations such as foreign NGOs and universities into not examining the situation here.

The next step is for the Papuan people, the Indonesian state and regional governments such as Australia's to acknowledge the irregularities in the Act of Free Choice and work to address this central question. Indonesian leaders need to stop basing their entire response to the West Papuan issue solely on the basis of the inviolability of the territorial integrity of Indonesia, territorial integrity that Evans and Jones point out is based on a fraudulent referendum.

John Martinkus
Jayapura, February 2003

Senator John Cherry is the Australian Democrats Senator for Queensland.

Greg Barns is an Australian Democrats member and former senior adviser to the Howard government.

Brian Coman is a former research biologist. As well as numerous scientific papers, he is the author of *Tooth and Nail: The Story of the Rabbit in Australia*.

Gareth Evans is the President of the International Crisis Group and former Foreign Minister of Australia, 1988–96.

Tim Flannery has made contributions of international significance to the fields of palaeontology, mammalogy and conservation. His books include *The Future Eaters*, *Throwim Way Leg* and *The Eternal Frontier*. He is the Director of the South Australian Museum.

Sidney Jones is Indonesia Project Director of the International Crisis Group, former Asia Director of Human Rights Watch and former Director of the Human Rights Office of the UN Transitional Administration in East Timor, 1999–2000.

William J. Lines is the author of *Taming the Great South Land: A History of the Conquest of Nature in Australia* and *A Long Walk in the Australian Bush*.

Ian Lowe is Emeritus Professor at Griffith University, where he was previously Head of the School of Science. He directed the Commission for the Future in 1988 and chaired the advisory council that produced the first national report on the state of the environment in 1996.

John Martinkus is an Australian investigative reporter on the Asia region. In 1999 he was nominated for a Walkley Award for his coverage of the violence in East Timor. His book *A Dirty Little War*, an eyewitness account of East Timor's struggle for independence, was shortlisted for the NSW Premier's Literary Awards in 2002.

Graham Richardson was federal Minister for the Environment from 1987 to 1990. His autobiography, *Whatever It Takes*, was published in 1994.

Paul Rogers has worked as an environmental consultant in chemical hazard assessment, including work on Australia's National Pollutant Inventory. He is the author of books about environmental issues in food and agriculture.

George Seddon is Emeritus Professor of Environmental Science (University of Melbourne). Among his books are *Landprints* and *From the Country*, an anthology from the work of T.R. Garnett, which he edited.

Jack Waterford is editor-in-chief of the *Canberra Times*.

QUARTERLY ESSAY

SUBSCRIPTIONS Receive a discount and never miss an issue. Mailed direct to your door. 1 year subscription (4 issues): $46.95 a year within Australia incl. GST (Institutional subs. $52.95). Outside Australia $74.95. All prices include postage and handling.

BACK ISSUES Please add $2.50 postage and handling to your order (or $8.00 for overseas orders).

- ☐ **Issue 1** ($9.95) Robert Manne's *In Denial: The Stolen Generations and the Right*
- ☐ **Issue 2** ($9.95) John Birmingham's *Appeasing Jakarta: Australia's Complicity in the East Timor Tragedy*
- ☐ **Issue 3** ($9.95) Guy Rundle's *The Opportunist: John Howard and the Triumph of Reaction*
- ☐ **Issue 4** ($9.95) Don Watson's *Rabbit Syndrome: Australia and America*
- ☐ **Issue 5** ($11.95) Mungo MacCallum's *Girt by Sea: Australia, the Refugees and the Politics of Fear*
- ☐ **Issue 6** ($11.95) John Button's *Beyond Belief: What Future for Labor?*
- ☐ **Issue 7** ($11.95) John Martinkus's *Paradise Betrayed: West Papua's Struggle for Independence*
- ☐ **Issue 8** ($11.95) Amanda Lohrey's *Groundswell: The Rise of the Greens*

PAYMENT DETAILS I enclose a cheque/money order made out to Schwartz Publishing Pty Ltd. Please debit my credit card (Mastercard, Visa Card or Bankcard accepted).

Card No. ☐☐☐☐☐☐☐☐☐☐☐☐☐☐☐☐☐☐☐

Expiry date   /   Amount $

Cardholder's name

Signature

Name

Address

Email

POST OR FAX TO:
Black Inc.
Level 5, 289 Flinders Lane, Melbourne,
Victoria 3000 Australia
Tel: 61 3 9654 2000 Fax: 61 3 9654 2290
Email: quarterlyessay@blackincbooks.com

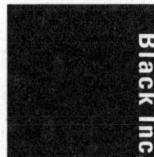

Black Inc.

Subscribe online at www.quarterlyessay.com